JOHN LINGENFELTER

ON MODIFYING SMALL-BLOCK CHEVY ENGINES

HIGH PERFORMANCE ENGINE BUILDING
AND TUNING FOR STREET AND RACING

HPBooks

HPBooks
are published by
The Berkley Publishing Group
200 Madison Avenue
New York, New York 10016

First Edition: February1996

© 1996 John Lingenfelter
10 9 8 7 6 5 4

Library of Congress Cataloging-in-Publication Data

Lingenfelter, John.
    John Lingenfelter on modifying small-block Chevy engines : high performance
engine building and tuning for street and racing / John Lingenfelter.
    — 1st ed.
        p.   cm.
    Includes index.
    ISBN 1-55788-238-X
        1. Chevrolet automobile—Motors—Modification.   2. Chevrolet
automobile—Performance.   I. Title
    TK215.C44L55   1996   95-25235
    629.25 '04—dc20   CIP

Book Design & Production by Bird Studios
Interior photos by the author unless otherwise noted
Cover photos by Randy Lorentzen

NOTICE: The information in this book is true and complete to the best of our knowledge. All recommendations on parts and procedures are made without any guarantees on the part of the author or The Berkley Publishing Group. Tampering with, altering, modifying or removing any emissions-control device is a violation of federal law. Author and publisher disclaim all liability incurred in connection with the use of this information.

CONTENTS

WHO IS JOHN LINGENFELTER?

The scene is classic Lingenfelter and one that has become part of drag racing legend. It's the 1976 U.S. Nationals, just north of Indianapolis, Indiana. For most drag racers, this is the race of the year. Everyone wants to win Indy. The thrash is on. John Lingenfelter has the Super Stock field covered in his Charlie Graf Enterprises-sponsored SS/LA Corvette until the engine breaks during the morning time trial the day of eliminations! The spare shortblock is back at John's shop in Decatur, Indiana, a round trip four-hour drive, which is out of the question. Unwilling to admit defeat, John discovers a helicopter nearby that, for a fee, will fly him to his shop, pick up the shortblock and fly him back in barely enough time to make first round.

Pandemonium erupts when John returns with the new engine. There are literally dozens of friends, fellow racers, and on-lookers contributing to the scene surrounding Lingenfelter's pit. The call for first round comes and the car's almost ready. Just a few more details like the fuel line, carb linkage, and ignition connections need completing. Work continues as they push the race car to staging.

The 'Vette is the last car in line as the thrash concludes, John straps in and spins the engine. It fires but sounds ragged. A last minute diagnosis fails to uncover the cause and an attempt at a burnout only confirms something is wrong. The valiant effort is over and John watches as his opponent solos for a first round victory. Back in the pits, a closer investigation reveals a chunk of silicone has clogged the carburetor and caused the problem. But the day is over.

What does this reveal about John Lingenfelter? Like most racers, John is a man who makes things happen. This attitude permeates not only his racing ventures but everything he faces in life. High performance engines and a desire to push them to unprecedented performance levels is a way of life with this man. Gain access to his inner circle of friends and you discover this soft-spoken engineer is never very far away from the scream of a small-block or the thundering torque of a big-block Chevy. While his pro football physique presents a commanding appearance, he shuns the limelight and is much happier twisting the controls of a dyno than as the center of attention in a crowd of well-wishers at a national event. He doesn't take time off for normal vacations like most people. Time away from the shop usually finds him drag racing at national events or diversions like racing at 200 mph in the Silver State Challenge. The rest of the time he's either yanking the handle on the dyno of

Lingenfelter's first NHRA national event win came at the wheel of this '69 SS/NA Camaro at the 1972 U.S. Nationals. For most drag racers, winning the U.S. Nationals just once is a lifetime goal. Lingenfelter would go on to win Indy four times. Photo: NHRA

anything from a fuel-injected 350 Corvette street engine or finessing his Pro Stock-inspired A/Econo Dragster engine.

Accomplished engine builders aren't born, they're created out of a ceaseless learning process. In John's case, his knowledge comes from an enviable combination of a strong engineering background combined with an innate hot rodder's curiosity. It's this quest that won't allow him to pass up the chance to try a new idea on the dyno and wring it out until it either succeeds or ends up as an expensive but dusty relic on an already-crowded shelf of prior experiments.

John Lingenfelter's automotive initiation dates back to his father's automotive skill as a technician in his hometown of East Freedom, Pennsylvania, just outside of Altoona. His hunger for anything automotive led him to Penn State University where he studied Mechanical Engineering. With

this sound engineering foundation, practical application began with a stint running the massive dynos for International Harvester. But beyond those thumping diesels was an intense desire to go drag racing.

No one builds a race car the first time out and ignites the drag racing fraternity. Nevertheless, his fellow racers did take notice when John won Super Stock eliminator at the U. S. Nationals at Indianapolis in a SS/NA '69 Camaro in 1972. It would become commonplace to find him in the winner's circle at an NHRA event. By '76, he was a fixture in Super Stock with the Graf Enterprises SS/LA Corvette, a car that John admits is still his favorite.

This quest for titles took him into then-Modified Eliminator with a short stint in 1977 with fellow-Indianan Bob Glidden's ex-Pro Stock Monza, the only Chevrolet that Glidden ever campaigned. Running

the car as an Econo Altered, Lingenfelter scored two national event victories and a runner-up finish. The consistency and ease of working on a dragster soon enticed John to switch to a B/Econo Dragster in 1978, producing his best quarter-mile year commanding three national event victories with three more runner-up finishes. A '79 Gatornationals win capped this streak but by then, John's fledgling engine building business began to demand more of his time.

Lingenfelter stayed away from active competition for six years but drag racing appealed to him again in 1985 when he fielded an A/Dragster that became the first six-second Competition Eliminator car that also won the Mile High Nationals along with three runner-up finishes. Then in '86, John scored his fourth U.S. Nationals title along with two more runner-up places. At the end of that year, John again took a sabbatical from drag

While Lingenfelter rarely looks back, this Charlie Graf-sponsored Corvette is still one of his favorites, winning the 1976 NHRA Fall Nationals and a runner-up at the Gatornationals. Photo: NHRA

John calls this the "ugly car". Beautiful it may not have been, but this B/Econo Dragster carried John to his most successful drag racing season with three NHRA Competition Eliminator titles and an equal number of runner-up finishes.
Photo: NHRA

racing to concentrate on his expanding business. This lasted until 1991 when he again entered the Competition Eliminator fray with a brand new A/Econo Dragster that also made history by becoming the first six-second Econo Dragster on its second outing. Competing on a limited basis at races close to home, John's perseverence paid off with a victory at the 1993 Western Auto Nationals in Topeka, Kansas, totalling an amazing 13 national event victories spanning 22 years.

During his two lengthy absences from drag racing, John merely redirected his energies into building a business that has become as successful as his engines are powerful. Expanding on his race engine business, John moved into the growing mail-order street engine business in 1984. Within a few months of beginning this venture, John was flooded with requests to apply that Lingenfelter magic to street engines, which set the course for John's most recent direction.

With the introduction in 1985 of Chevrolet's electronic fuel injection Tuned Port Injection (TPI) system in the

This B/Econo Altered was the ex-Bob Glidden Pro Stocker that Lingenfelter campaigned in Competition Eliminator in 1977, posting two national event wins and a runner-up. Photo: NHRA

For his return to Competition Eliminator in 1985, Lingenfelter built this A/Dragster that was the first Competition Eliminator car in the 6's, winning the Mile High Nationals along with three runner-up finishes that year. Photo: NHRA

Corvette, Lingenfelter realized the tremendous opportunities that the TPI system promised. Since then, Lingenfelter has raised the level of power produced by these multi-point fuel injection systems to unprecedented levels. While he has numerous competitors, Lingenfelter Performance Engineering is the unquestioned leader in creating amazing power from these electronic small-blocks.

But drag racing is only one part of what makes up this man. Few enthusiasts are aware of Lingenfelter's other varied automotive interests. While he acknowledges no grand design, his experience with a variety of performance applications has also expanded his knowledge base.

One of his most famous ventures outside the quarter-mile world was a

If judged strictly on speed, this is Lingenfelter's greatest achievement. The Firebird unofficially hit 298 mph at Bonneville, but never attained the official two-way 300 mph record. While falling short of his goal to drive the first production sedan over 300 mph, this 298 mph Firebird still stands as an awesome achievement.

Lingenfelter's black '94 383 EFI Pontiac Trans-Am appeared at the '94 SEMA show to illustrate how it's possible to build an emissons-legal, 12-second street car that combines low emissions with awesome torque and respectable fuel economy.

collaboration with Reeves Callaway to develop the engine for the Sledgehammer Corvette in 1988. Lingenfelter built two streetable twin-turbocharged 355-inch small-blocks for the car, with the final engine producing an honest 900 horsepower! Once the car was configured, Callaway's employees drove the Corvette from Callaway's shop in Connecticut to the Transportation Research Center (TRC), a 7.5-mile oval in East Liberty, Ohio. TRC's steeply banked turns were an ideal venue for a closed-circuit, top speed attempt. On test day, Lingenfelter volunteered to drive the car and after some innovative last-minute tweaks, the 'Vette screamed to an unbelievable 254.76 mph top-end blast at 6200 rpm in 5th gear! After the record-breaking run, the Sledgehammer was then driven back to Connecticut, a testimony not only to Lingenfelter power but also to its durability.

Later invitations to attend top speed shootouts have only bolstered that reputation. In 1990, John configured his red 1986 Corvette test car with full safety equipment and built a thumper 408 cid small-block cranking out 540 horsepower at an amazingly low 5400 rpm to compete in the Nevada Silver State Classic top speed contest. Despite melting both the TH700-R4 automatic and the Gear Vendor's overdrive, the 'Vette posted a 157 mph average for the 93-mile course while sustaining a speed over 206 mph for over five minutes! In classic Lingenfelter form, John was disappointed in the performance. Placing third overall, he thought it should have run much faster!

John's greatest top speed effort, and perhaps his most frustrating, took place on the barren stretch of salt at Bonneville. One of the last great frontiers at the salt flats is breaking the 300 mph barrier in a production sedan. John's experience with the Sledgehammer Corvette led him to believe that 300 mph in a slightly more aerodynamic car like the Firebird was

This is Lingenfelter's latest race car, a Chassis Craft A/Econo Dragster that most recently won the '93 Topeka, Kansas, NHRA Heartland Nationals. John also set a national record with this car as the first A/Econo Dragster in the sixes. Photo: NHRA

possible. Pushing a 355ci small-block to its ultimate potential, Lingenfelter was able to squeeze an incredible 1400 horsepower from this twin-turbocharged Mouse motor, carefully balancing outrageous power with reliability. Since intercoolers are essential with boost levels of this magnitude, John employed a bank of six huge nitrous bottles under the Firebird's rear hatch. But he didn't use them the way you might think. Rather than inject the nitrous into the engine, John cunningly used the gas as a cooling medium for the on-board, air-to-air intercoolers. Introducing the highly pressurized nitrous into the intercoolers radically dropped the turbochargers' outlet temperatures, which does wonderful things for horsepower.

The first attempt began in October 1989 with partners Carl Staggemeir and Gary Eaker. With Eaker driving, the car hinted at its potential when the car hit 298 mph on an early down run averaging 293 mph with a backup run the following morning. A bizarre, high-speed spin prevented any further attempts that year. The following year, Lingenfelter returned but failed to attain the car's previous performance marks due to a variety of problems. Because of mounting costs and no sponsorship, the effort was eventually shelved. Despite missing the 300 mph record, it's undeniable that when it comes to powerful small-block Chevys, Lingenfelter has the key.

From all these experiences, Lingenfelter has compiled a tremendous data base that is invested in virtually every engine that leaves his shop.

Lingenfelter Performance Engineering has built engines for circle track cars, numerous drag racing classes, motor homes, off-road trucks, national champion SCCA autocross racers and even powerplants for law enforcement vehicles and ambulances! Taking these experiences as a whole, there are few small-block performance paths that John Lingenfelter has not tested, tried, and bested for the past 20 years. It is this on-going experience and John's ceaseless quest to continually improve existing power levels of the high performance street small-block Chevy that this book will attempt to cover.—*Michael Lutfy*

For more information, contact: Lingenfelter Performance Engineering, 1557 Winchester, Decatur, IN 46733. 219/724-2552. ∎

BASIC ENGINE THEORY 1

The recipe for building a stout small-block Chevy might seem relatively simple. Start with a good short block, stuff in a healthy camshaft, bolt on a set of hogged-out cylinder heads, pop on an intake, carb and headers and there you have it—an award winning small-block. If you've ever tried to build an engine with little more preparation than this, then you probably know that nothing could be further from the truth. While most hot rodders are often overly enthusiastic about building their engine and want to dive right into the project, I have found that some important questions need to be answered long before the first parts are purchased or machine operations begin.

This chapter is intended to give you a theoretical perspective on the dozens of details that combine to create a particular engine power curve. This will be a quick overview of basic engine theory and also how the vehicle and drivetrain interact to dictate how engine power will make the car perform. Don't be too concerned if after reading this chapter you still don't feel like you have a grasp on all the details. Many racers have spent the better part of their lives creating, learning, and massaging this theory in order to better maximize acceleration, so don't feel that you must be on a par with them after only one quick trip through this book.

There are many details that are combined to create a particular engine power curve. This chapter is to give you a theoretical overview of basic engine theory and how the engine and drivetrain interact to produce power.

STREET POWER

Before we get into the heart of engine design, it's worth mentioning the classic concept of "there is no replacement for displacement." Most enthusiasts start with what they already own. However, if your sole small-block is a paltry 283 or 307, it will be challenged to perform as well as a larger 350 or 400 cubic-inch engine. Adding displacement is the quickest, easiest, and often least expensive way to increase power throughout the entire rpm band, especially in the lower rpm range where torque is king. As you will see throughout this discussion, torque plays a major role in street car acceleration.

Application

The first point that needs to be addressed is the specific application of the engine. Since this book is targeted at street performance, that will be the goal of most of the engines described here. However, a potent street small-block is a very broad category, depending upon

Building torque is a great way to accelerate even a heavy Camaro in tight autocross or road racing applications. A great combination would be a 2800 pound car with 500 lbs-ft of torque and enough traction to propel it out of the corners.

What you feel when you hit the throttle isn't horsepower, it's torque. Torque is what accelerates a car. In fact, horsepower is merely torque over time (rpm). Increasing torque in a streetable rpm range, between 2500 and 4500, will turn a lazy street car into a stormer.

THE LINGENFELTER POWER FORMULA

• Always maximize power within the rpm band where the engine spends its most time.

• Vehicle application is far more important than most enthusiasts would believe.

• Bigger is rarely better. "Small" intake ports, manifolds, and exhaust systems maximize torque which improves that wonderful "seat-of-the-pants" feeling in street engines.

• Port velocity will improve cylinder filling more efficiently, especially at and below peak torque, than big flow numbers generated by large, slow-moving ports.

• A strong cylinder head will always make more power even if poorly matched with the cam. A weak cylinder head matched with an optimized cam will never perform as well.

• Transmission and rear end gear ratios have a significant impact on engine component decisions. A good example of this is camshaft selection charts that emphasize cruising rpm—basing, in part, a cam profile on gear ratios.

whose definition you believe. As you will see, the application of your street small-block can become somewhat complex. Is this engine your only means of transportation? Will it be primarily street driven or bracket raced? If you decide that the engine is to be primarily street

driven as daily transportation, it would be foolish to specify a 12.5:1 compression ratio requiring 110 octane race gas. These are contradictory requirements for any engine.

Once the basic usage questions have been defined, only then can we move on

to the vehicle in which the engine will be used. Many engine builders will construct a very powerful engine intended for some vague "performance" application. Some of the better engine builders will even tell you that you will have to make some modifications to your car in order to take

POWER VS. ACCELERATION

GRAPH 2-1

This graph shows the time spent in each of three rpm bands. As you can see by the graph, this small-block spends 51 percent of its time in the 4000 to 5000 rpm band, more than the other two bands combined. For this Camaro, concentrating on improving power in the 4000 to 5000 rpm band would pay off in better acceleration. The next best place to concentrate efforts would probably be in the 5000 to 6000 rpm range as long as it does not detract from mid-range power. This is based on the assumption that the car will remain unchanged. Altering rear gear ratio, changing transmission gear ratios or other major changes would require data logging another rpm curve. This is illustrated in graph 2-2.

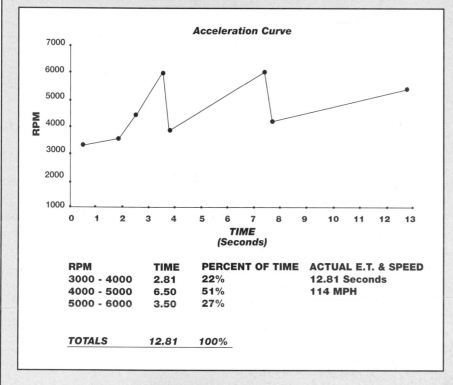

Acceleration Curve

RPM	TIME	PERCENT OF TIME	ACTUAL E.T. & SPEED
3000 - 4000	2.81	22%	12.81 Seconds
4000 - 5000	6.50	51%	114 MPH
5000 - 6000	3.50	27%	
TOTALS	**12.81**	**100%**	

advantage of your new-found power. While all of this is unfortunately true, very few take the time to build a specific package that will suit your particular vehicle and application. This leads us to the first and most critical part of what I call my "Lingenfelter Power Formula."

The Golden Power Rule

The foremost rule (to which there seem to be no exceptions) is to always emphasize power within the rpm band where the engine spends a majority of its time. Think about this for a minute. This is the single most important concept you can learn from this book! As an example,

let's take a typical street-driven Camaro with a 383 small-block, Turbo 400 automatic transmission, 3.50 rear gear ratio, and weighing 3300 lbs with a driver, ready to run. Let's make a full-throttle run down the drag strip while recording engine rpm for the entire run. The material we'll use for this example was generated using the RPM Analyzer software from Performance Trends and a laptop computer. In graph 2-1, we have outlined just such an rpm trace of a mid-12 second small-block Chevy. By recording data every .10 second, we now have a way to determine the amount of time the engine spends in any particular

rpm range.

To simplify this exercise, we've broken the entire rpm band into three categories: low-to-mid rpm, mid-to-high rpm and the peak horsepower band. Automatic transmissions have become very popular, especially in performance drag racing applications since they make the car easier to drive and reduce driveline parts breakage. However, because of the wide spread between the three forward gears, automatics suffer from a wide rpm drop between gear changes. For example, a Turbo 400 automatic offers stock ratios of 2.48/1.48/1.00 which means the rpm drop between first and second gears at a 6000 rpm peak will pull the engine down to approximately 400 to 500 rpm above the stall speed of the torque converter, around 3900 rpm using the above Camaro as an example. This means the engine must have the capability to operate in at least a powerband of 2100 rpm. But there's more to it than just this simple math.

The first example is the high 12-second Camaro described above. In Graph 2-1, we've depicted its acceleration curve. I prefer to define the bottom of the lowest rpm band as 400 to 500 rpm below the stall speed of the torque converter as given above. With a 3400 rpm stall speed, the low-to-mid rpm band would span from 3000 rpm to 4000, the mid-to-high band from 4000 to 5000 rpm, and finally the peak horsepower band of 5000 to 6000 rpm.

By recording the amount of time the engine spends in each rpm band, we create a "residence time" in the low-to-mid rpm band of 2.81 seconds, a time of 6.5 seconds for the middle band, and 3.5 seconds of time spent at the higher rpm levels. This simple test makes it obvious where the engine spends more than half of its time. While the engine does spend over a quarter of its time at the higher rpm levels, this exercise is intended to point out the error in concentrating only on peak horsepower levels.

If you were given the choice of where

GRAPH 2-2

This graph illustrates an 11-second car with a 4-speed and more power. The additional gear in the transmission changes the acceleration curve so that improving power in the middle band from 5000 to 6000 rpm would be the most beneficial. The next place to work on improving power would be in the 4000 to 5000 rpm band. Notice how the addition of another gear limits the rpm drop keeping the engine in a narrower rpm band. The shape of this curve is also steeper than the automatic's curve.

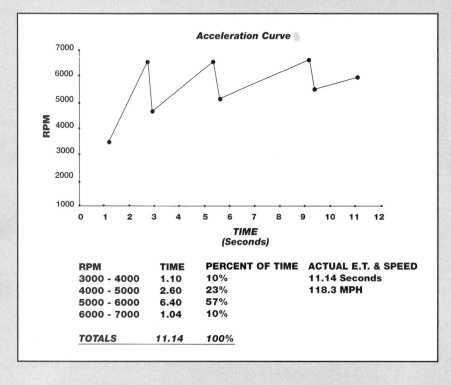

Acceleration Curve

RPM	TIME	PERCENT OF TIME	ACTUAL E.T. & SPEED
3000 - 4000	1.10	10%	11.14 Seconds
4000 - 5000	2.60	23%	118.3 MPH
5000 - 6000	6.40	57%	
6000 - 7000	1.04	10%	
TOTALS	**11.14**	**100%**	

Whether it's a brand new Corvette, Firebird or '55 Chevy, the theory remains the same. Decide how the car will be used and then build the engine to fit that application. This is usually much less expensive than building an engine that requires a complete reconfiguration of the car.

you would want to optimize power for this particular vehicle, where would you place your emphasis? If you said the area where the residence time is the greatest, give yourself one gold star.

Graph 2-2 illustrates an rpm band trace of an 11-second, 4-speed manual transmission-equipped small-block that reveals a more aggressive rpm trace. The addition of one extra gear in the transmission tightens the rpm drop between the gear changes, allowing the engine builder to concentrate his power emphasis on a narrower rpm band. Since this engine operates at a higher rpm level, we've also added another time band from 6000 to 7000 rpm. The resulting graph reveals an engine that spends over half its time in the 5000 to 6000 rpm band. Here, the trend toward additional time spent in the mid-to-high rpm illustrates that this manual trans application would respond to tuning the engine in this rpm band.

Assuming that 3-speed automatics require a wider rpm power band than 4-speed manual transmission-equipped cars, this concept dictates the way the engine should be built. For example, I specify my Lingenfelter/ACCEL SuperRam intake with its longer intake runners for a TH700-R4-equipped Camaro or Corvette, but if that same car is equipped with a 4-, 5- or 6-speed manual transmission, then I usually choose the shorter runner length Chevrolet LT-1 intake manifold. If you've guessed that intake manifold runner length has some effect on the engine's torque curve, give yourself another gold star.

INTAKE MANIFOLD TUNING

Engineers call intake manifold size and length the "organ pipe" concept of intake manifold tuning. This is a constant, regardless of whether the engine utilizes a carburetor or electronic fuel injection (EFI). This relates to how a musical instrument like a trombone or calliope

4

Intake manifold tuned length plays a big part in determining both the amount of torque the engine can make and where, in the rpm band, this power will occur. Long runner length manifolds such as the TPI intake (right) make great torque but at the sacrifice of high rpm power. The Lingenfelter/ACCEL SuperRam intake (center) is a great compromise between the TPI intake and a short runner intake such as the new LT1 intake (left).

The SuperRam intake makes outstanding torque, which is why it's so popular. This is the late Mark Thorton's black Corvette with its stout Lingenfelter 406 small-block. This car competed at both the Automotive Triathlon and the Silver State road race in Nevada.

changes pitch by changing the length of the path the air must travel. In fact, intake runner length is one of the critical decisions in engine building since it contributes significantly to the shape of the power curve.

Intake Runners

Taking this idea one step further, many feel that electronic fuel injection, by itself, is responsible for dramatic power increases. This, unfortunately, is not entirely true. There are basically two

variables in the design of an intake system—intake runner diameter (often called cross-sectional area) and the length of the runner.

What makes a stock Chevy 350 Tuned Port Injection (TPI) engine produce impressive mid-range torque is the added runner length of the intake manifold, not because the engine is equipped with individual fuel injectors. Nor is the concept of long runner lengths a new one. Chrysler built an extremely long intake runner, dual four-barrel, cross-ram carbureted big-block back in the early '60s that made exceptional torque but was hampered by cold-start problems because the carburetors were located such a long distance from the intake manifold heat that helps vaporize fuel in cold climates.

Attention to runner length applies to carbureted intake manifolds as well. Dual plane intakes, such as the Edelbrock Performer and Performer RPM, are designed to increase runner length over single plane intakes, such as the Victor Jr. or Holley Strip Dominator. Enthusiasts often ask which style of intake is better than the other. As you can see from our investigation of the required rpm bands of the above Camaro example, the choice of intakes should be dictated by your emphasis on a specific rpm powerband for the engine. If you require power in the lower to mid-rpm bands, then a dual-plane intake would be a wise choice. If the engine spends more time in the upper powerbands (as in a 4-speed application), then a single-plane intake could offer improved elapsed time (e.t.) potential. Notice how virtually every decision made about the car, from the gear ratios to transmission choice, has an effect on the engine application. Think of the entire car as an integrated system in which all the hundreds of parts must work together in order for the car to perform to its maximum potential.

Condensing this very complex subject down to its basic concepts, increasing runner length tends to improve engine

TORQUE CURVE TUNING

This graph clearly illustrates the effect of runner length on the torque curve. The Lingenfelter-designed ACCEL SuperRam intake utilizes a significantly longer runner compared to the LT1 short length intake, both on a Lingenfelter 383. With everything else remaining the same, note how the SuperRam pumps the torque curve in the mid-range while the LT1 intake is stronger above 5500 rpm.

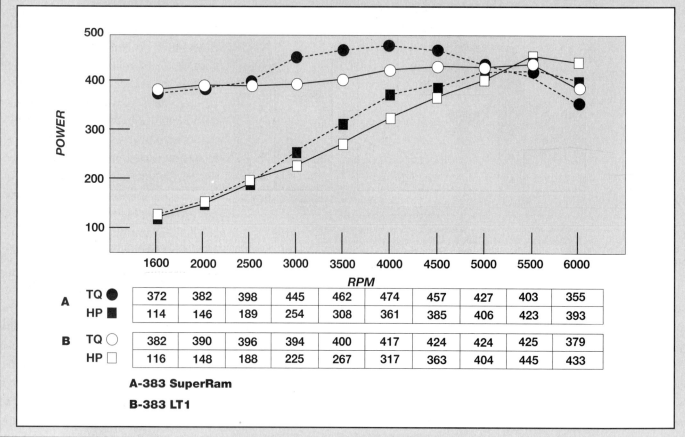

			1600	2000	2500	3000	3500	4000	4500	5000	5500	6000
A	TQ ●		372	382	398	445	462	474	457	427	403	355
	HP ■		114	146	189	254	308	361	385	406	423	393
B	TQ ○		382	390	396	394	400	417	424	424	425	379
	HP □		116	148	188	225	267	317	363	404	445	433

A-383 SuperRam

B-383 LT1

torque at the lower rpm levels while simultaneously lowering the peak torque rpm point. Since there are very few free lunches in this world, the downside to increasing low rpm torque is a subsequent loss of horsepower occurring usually at a lower peak rpm. At the other end of the spectrum, an extremely short runner intake (such as a Victor Jr. single plane carbureted intake) offers improved power levels at higher rpm levels, sacrificing power usually below peak torque in favor of the higher rpm levels.

Cross-Sectional Area—The other half of the intake manifold design process, cross-sectional area of the port, also plays an important role. Generally, a smaller runner area increases intake

charge velocity, speeding up the inlet charge, which improves cylinder filling (expressed as volumetric efficiency) at lower rpm levels. Conversely, huge intake manifold port runners contribute to slowing the intake gas speed at lower rpm levels, hurting power below peak torque while contributing to improved cylinder filling at rpm levels closer to peak horsepower.

It's important to remember that horsepower is merely a function of torque at a higher rpm where there is less time to fill the cylinder. If the engine makes acceptable torque at higher rpm, it will always make good power. The key is configuring the engine to flow vast amounts of air in a very short time. For

example, at 6000 rpm, the engine must fill the cylinder 50 times per second. As you can imagine, that's not much time to do the job!

TPI Corvette Intake—A classic example of a long runner length, small cross-sectional intake emphasizing low rpm torque is the '86 to '92 L-98 TPI Corvette engine. With its exceptionally long runner length intake manifold, this engine makes great torque in its stock configuration because the small runners contribute to high inlet air speeds even at low engine speeds. This high gas speed at the lower engine speeds fill the cylinder with greater amounts of air and fuel. This is why the engine makes great torque. On the down side, this manifold suffers from

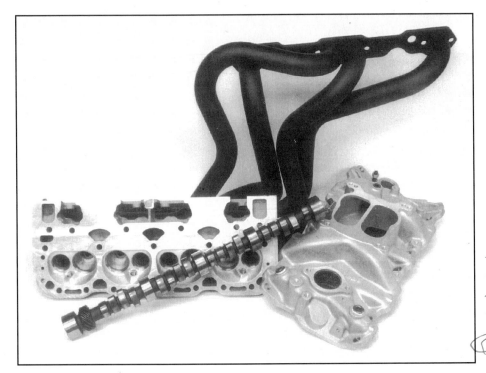

The careful matching of head port flow, camshaft timing, intake manifold design and header/exhaust system are the basic components to building a strong street engine. One mismatch or incorrectly chosen component can hurt the overall power potential of the engine.

a peak horsepower rpm of barely 4800 rpm. This low peak rpm horsepower point is a combination of the extremely long runner length and small cross-sectional area of the intake port runners. Not only does this small intake port area restrict flow, but the air must travel a greater distance from the intake plenum to the intake valve. That means less air is able to travel the greater distance, restricting the ability of the engine to make power at those higher rpm levels.

On the other side of the coin, a short, large cross-sectional intake port, such as a big Edelbrock Victor or Holley Strip Dominator intake, usually suffers from "slow" intake gas speed at engine speeds below peak torque because of the large port diameter and the lack of column inertia that is generated by a long runner length port. These manifolds usually excel at making power above peak torque where time to fill the cylinders is limited. The short runner length does not restrict the flow as much and the larger port area now contributes to higher gas speeds at this higher rpm.

COMBINATIONS

While it's beyond the scope of this chapter to go into the camshaft and cylinder heads, they also play an integral part in the overall power equation.

Heads & Cams

My approach has always been to maximize torque for street-driven engines, which means avoiding the large cross-sectional area cylinder heads in favor of smaller intake runner cylinder heads with excellent flow numbers. Typically, this also means choosing a cylinder head with good intake flow and an exhaust port with an exhaust-to-intake relationship of at least 70 percent. This means that at around .450-inch lift, the exhaust port flow cfm is 70 percent of the intake port flow cfm. For example, a small-block Chevy head with 240 cfm flow at .450-inch valve lift should have an exhaust port capable of flowing at least 168 cfm. Ideally, exhaust-to-intake percentages of closer to 80 percent are even better, especially with a good intake port.

Camshaft Design—Camshaft design must also be integrated with the cylinder heads, intake, and exhaust systems. Since camshafts are relatively inexpensive compared to cylinder heads and other internal engine pieces, enthusiasts often go overboard in camshaft selection, suffering from the "more is better" syndrome. Maximizing the engine combination for a particular rpm band means choosing the appropriate camshaft as well. Long duration, high lift cams emphasize power in the upper rpm ranges while shorter duration camshafts pump the torque curve at some sacrifice of upper rpm horsepower. Usually, the choice of a small intake port volume of between 170 to 180cc's is compatible with a short duration flat-tappet hydraulic camshaft in the neighborhood of 220 degrees of seat duration at the .050-inch tappet checking height.

By integrating the two separate intake and exhaust systems (along with the obvious inputs from the camshaft and cylinder heads) you create a particular power curve. Like the intake tract, the exhaust system should be viewed as a whole. Since this book is geared toward the street enthusiast, the entire exhaust includes headers, exhaust pipes, mufflers and tail pipes. Enthusiasts often build a strong engine only to strangle it with a small, inefficient exhaust system. This is a quick way to kill power.

Exhaust

You can think of the exhaust system as similar to a backwards intake tract. The primary header tubes are similar to intake runners, while the header collector is not much different than the plenum of the intake manifold. With that in mind, it would make sense that choosing headers would be somewhat similar to choosing an intake manifold. Therefore, increasing header primary tube diameter (1 3/4-inch for a 350 for example) would tend to enhance power in the upper rpm ranges while perhaps hurting power slightly

7

The search for power doesn't end once the basic components are selected and combined. Dyno testing and in-car testing are still required to produce the optimal timing and jetting. Dyno testing is helpful but actual vehicle testing is the only true way to optimize these variables.

below peak torque. Conversely, the typical 1-5/8-inch primary pipe header would enhance power at a lower rpm street-oriented peak torque while contributing less to peak horsepower. As you can imagine, primary exhaust pipe length also has an effect on power in much the same way as the intake tract affects power. Typical small-block street headers generally fall in the range of 32 to 38 inches in primary pipe length.

Header Collector Volume— Viewed like an intake manifold plenum, exhaust header collector volume has been shown to deliver measurable increases in torque without the subsequent losses in opposite rpm power. For example, increasing the length of a header collector often improves power below peak torque without sacrificing high rpm horsepower. In terms of street exhaust system design, increases in exhaust "collector" volume can be achieved by the use of an exhaust cross-over or H-pipe that adds additional exhaust system volume without restricting exhaust flow. The proper size and positioning of an H-pipe has often proved to be not only worth significant increases in power but also contributing to sound level reductions as well.

As you can see, there are hundreds of variables that affect the decisions that must be made to optimize power for a

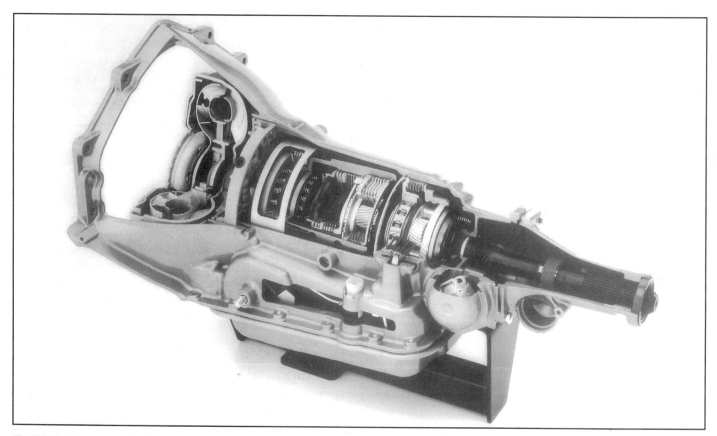

The 700R-4 is a great overdrive automatic, but the deep 3.06 first gear makes the 1-2 and 2-3 shift rpm drops excessive. While the deep first gear ratio may help the launch, the gear change rpm drops hurt the overall e.t.

TRANS-ENDENTAL MEDITATION

Another way to look at the impact of transmission gear ratios is by envisioning them on a horizontal bar graph. The deeper the first gear ratio, the wider the engine's powerband must be to accommodate the rpm drop between shifts. The deep first gear ratio in the 700-R4 appears to offer an advantage of not requiring as much rear gear. Unfortunately, this 3.06 first gear ratio means the engine will experience a greater rpm drop between the 1-2 shift if compared to a Turbo 350 or 400 trans with a first gear ratio of roughly 2.50:1 since the 700-R4 second gear and the 350/400 transmissions have similar second gear ratios.

The Powerglide might appear to be a good idea, but its higher 1.82:1 first gear ratio doesn't produce sufficient gear multiplication to get a typical street car moving (unless you put an outrageously deep gear in the rear end) and there is a significant rpm drop when making the shift from low to high gear. This transmission works best in high torque, high horsepower drag racing applications, especially where a three-speed's deeper first gear may contribute to tire spin. The best compromise for most street cars is the TH-400 or TH-350 transmission, which is why, combined with their excellent durability, they are so popular. A good compromise is the 200-4R automatic if you are looking for a durable overdrive automatic since the 200-4R's first gear ratio is only 2.76:1. This same concept can also be applied to choosing the proper gear spreads for manual 4, 5 or 6-speed transmissions.

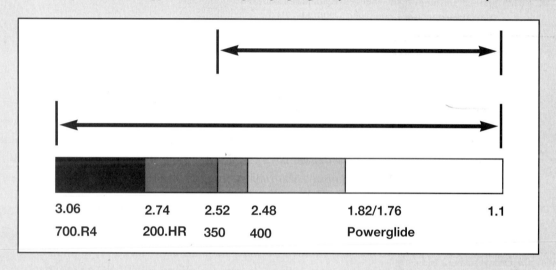

3.06	2.74	2.52	2.48	1.82/1.76	1.1
700.R4	200.HR	350	400	Powerglide	

Gear Ratios of Popular GM Automatics

	Powerglide	TH350	TH400	700R-4*	200-4R	4L80-E
1st	1.76/1.82	2.52	2.48	3.06	2.74	2.48
2nd	-	1.52	1.48	1.63	1.57	1.48
3rd	1.00	1.00	1.00	1.00	1.00	1.00
4th	-	-	-	.70	.67	.75

*The 4L60 and 4L60-E are the new designation for the 700-R4. The "E" signifies an electronically controlled transmission. The 4L80-E is the new "electric" Turbo 400.

street-driven engine. Cost, the vehicle's intended purpose, the amount of street miles it will be driven, fuel quality, fuel mileage, idle quality, vehicle weight, gear ratios, and rpm "residence time" plus a few hundred other important details all affect the choice of engine components since it is these pieces that will dictate the powerband the engine creates.

No one expects you to learn all the nuances of this theory in one sitting. By design, I have moved very fast and glossed over a number of important points. In later chapters I will expand on these basic concepts in more detail that will fill in some of the fuzzy corners that

I have purposely created to present the overall picture. Don't be afraid to go back and re-read this chapter in order to grasp the basic concepts. This will be helpful since the foundation laid in this chapter will help as I explain the secrets to building the powerful and efficient small-block Chevy of the future! ■

2 DISPLACEMENT COMBINATIONS

There are exceptions to any rule and perhaps when it comes to engine displacement, this is one place where "bigger is better" actually holds true. When it comes to making power, displacement is the easiest way to get there, especially when it comes to street engines where torque is king.

Bore Center & Deck Height

The small-block Chevy has appeared in a staggering number of production bore and stroke configurations since its inception in 1955. In stock trim, the Mouse has appeared in nine different bore and stroke combinations in its original envelope that includes *bore center* and cylinder *block deck height*. These two configurations are the two most important considerations for an engine designer. Bore center is the spacing between the bores, which limits the overall bore size. For the small-block Chevy, this is 4.4 inches between the centerline of each cylinder. This limitation is why the later 400 small-blocks came from the factory with siamesed cylinders, eliminating the water jacket between the bores in order to maintain sufficient cylinder wall thickness.

The other limitation of any V-type engine is block deck height. The dimension is measured as the distance from the crankshaft centerline to the top

Killer 420 cid small-blocks that make 500 horsepower are a long way from the original 265 cid small-block from 1955.

of the cylinder block at the head mating surface. The small-block's original deck height is 9.025 inches. The combination of one-half crankshaft stroke, connecting rod length and piston compression height must equal a figure no greater than the deck height of the block or the piston will protrude above the block height. While this information isn't critical when

assembling a stock engine, it's always good to know. Paying attention to these variables can be helpful in certain cases when the engine builder is mixing and matching small-block components. The classic example is when an inexperienced engine builder mistakenly added 4-inch bore 350 pistons to a small-block equipped with a 327 crankshaft. The

COMPUTING DISPLACEMENT

Displacement in this book is expressed as cubic inches. Just in case you were working on your car the day your geometry class learned how to compute the volume of a cylinder, or you're like us and math class was a long time ago, we've included this quick refresher course.

For our example, let's do a stock bore and stroke 350. We'll also use a constant to keep the math easy. The figure .7854 is pi divided by 4.

DISPLACEMENT = .7854 x (bore x bore) x stroke x number of cylinders
DISPLACEMENT = .7854 x (4.00 x 4.00) x 3.48 x 8
DISPLACEMENT = 349.849 rounded to 350 cubic inches

The 350 is the most prolific small-block bore/stroke combination with its 4.00-inch bore and 3.48-inch stroke. This engine is popular mostly because there have been more 350's built than any other small-block. You can build a 350 with stock cast pistons and a cast crank all the way up through a killer 4340 steel crank with NASCAR style rods and lightweight pistons.

engine ran, but was down on power. A subsequent tear-down revealed the pistons never came close to reaching Top Dead Center (TDC) because of the shorter compression height pistons that were designed for a longer 3.48-inch stroke crank than the 327's 3.25-inch stroke. The engine ended up with something like a 7.0:1 compression ratio because of this oversight, which is why it made no power.

In this chapter, we'll take a look at the more popular engine displacement combinations for street and mild competition Mouse motors. Between the factory displacement combinations and the plethora of aftermarket combos, this displacement discussion will be limited to five of the most popular bore/stroke combinations. We'll also take a look at a couple of smaller displacement engines that offer certain advantages if you are interested in combining performance with fuel economy. In addition, we'll give you some estimates of the power you could extract from a pump gas engine of that displacement. We'll go over actual power levels in more detail in Chapter 18 on dyno-tested combinations, but these estimates are based on engines that I have built that represent attainable power levels without using exotic roller cams or expensive ported heads.

350/355

It's appropriate that we start with the smallest yet most popular small-block of the family. Built since 1967, there are literally millions of these engines in virtually any type of vehicle that uses a small-block for power. With its relatively short 3.48-inch stroke and large 4.00-inch bore, the engine is a great compromise between bore and stroke for great overall power.

Because of this great bore/stroke relationship, the 350 fills the bill for either a basic production-based street engine or as a 7000 rpm endurance engine. This tremendous popularity also guarantees that specific parts like cranks, pistons, rings, bearings and other specific parts are priced competitively because of the sheer demand. Compare the price of the 350's 4.00- and 4.030-inch bore pistons against other sizes and you quickly see these are often the least expensive because of the tremendous volume this engine commands.

One of the other things we'll examine with these bore and stroke comparisons is the rod length to stroke ratios (R/L). This is explained more fully in Chapter 5, but this relationship is expressed as a ratio of the engine's rod length divided by its stroke. The 350's R/L ratio computes out to 1.64:1. As you will see, as the stroke increases it's necessary to lengthen the

rod in order to maintain the R/L ratio. An excellent ratio would be over 1.7:1. A 3.48-inch stroke with a 6.0-inch rod creates a 1.72:1 R/L ratio.

With a good cylinder head and a mild hydraulic flat-tappet cam of around 220 degrees of duration at .050-inch and a high rise dual plane like the Edelbrock Performer RPM, it's possible to make 420 lbs-ft. of torque and 400 horsepower.

The 372/377 looks interesting on the surface, but requires spacer bearings to use a stock stroke crank in the larger bore 400 block. The larger bore helps unshroud the valves, but for a street engine, the added expense doesn't justify the power you will make.

372/377

One small-block derivative you may have heard about is the 372/377 combo. The theme behind this engine is that a large bore helps unshroud the valves to improve breathing, especially at high rpm. The best way to accomplish this task is to use a 400 block's 4.125-inch bore but drop in the 350's 3.48-inch stroke crank. In essence, this combo is a destroked 400. Rather than buy an expensive forged steel crank, many budget racers use a stock, cast 350 crank. Since the 400 block requires a large 2.65-inch main journal and the 350 crank is a 2.45-inch main journal, spacer bearings are normally used.

Drawbacks—While it sounds like this would be a real screamer, there are concerns you should consider. First, cast cranks are not generally compatible with high rpm applications. A forged steel crank is a better choice. The cast crank must use spacer bearings that can create problems if not installed with the proper clearances. To eliminate the spacer bearings in the mains, you'll need a custom steel crankshaft, which is also

expensive. While you're at it, you might as well increase the rod length given the short stroke. As you can see, you're quickly speeding toward an expensive engine.

Perhaps the most important question is why you would want to invest so much money in an engine that automatically gives up cubic inches to a 383, 406 or 420 cid engine that would cost the same or less. Basically, this is a combination best chosen with rules that limit maximum cubic inches and where cost is not a major concern. Neither of these considerations are realistic for street engine builders.

383/388

This bore stroke combination is easily the most popular size for street enthusiasts. Its simplicity is only enhanced by the relatively low cost penalty for the added cubic inches. Assuming forged pistons are part of the plan, these added inches are a low-cost power plan. A cast 400 crank is the most popular choice to increase the stroke bolted into the 350 block. The 383 is

LET'S GET SMALL

While this chapter focuses on displacements of 350 cid and larger, there is a place for the small cubic inch small-block. The most populous of the small-inch Mouse motors is the 305 and its older cousin the 307. These engines are completely different bore and stroke combinations but offer the advantage of acceptable power with good fuel economy. Most hot rodders use these engines because that's all they have. Unless you are forced by economics to use these engines, we'd suggest stepping up to the 350. But if your bank account is on a serious diet, these small-inch engines will power your streeter just fine.

The 305 combines the 350's 3.48-inch stroke with a small 3.736-inch bore. Chevy did this to consolidate its two small-blocks with the same stroke. Both the 307 and 305 will respond to the same kind of basic modifications as any other small-block. The 305 offers some off-idle advantages in part-throttle power due to its longer stroke. Unfortunately, the 305 also seems to be overly-sensitive to detonation. Even with aluminum heads, the 305 will not tolerate excessive initial timing without detonating.

The larger bore, shorter stroke 307 is older than the 305 but offers perhaps a slight advantage with its larger bore. Interchangeability is complete just as with all small-blocks for heads, cams, intakes and exhaust. In any case, remember that since these engines are short on size, they should be used as efficiency leaders. This means conservative cam timing and compression using a dual plane intake and a small carburetor. This will create a crisp small-block that will reward you with thousands of miles of dependable power.

On the other hand, the longer stroke 383 is second only to the 350 in sheer popularity. The longer stroke 3.75-inch stroke pumps the displacement another 28 cubic inches which does wonders for torque. But best of all is the fact that no exotic parts are required since you can use a stock crank and 5.565-inch 400 rods if you're on a tight budget.

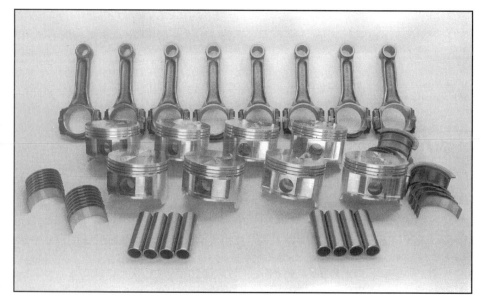

If you're searching for massive torque on a small-block budget, the 400 is the only logical choice. The cast crank is more durable than you might think and will handle power at 6000 rpm with no sweat. Forged pistons are a must.

based on a 4.030-inch bore while the 388 pumps the bore another .030-inch to 4.060-inch. The only specialty machining required is grinding the crank down to the 350's 2.45-inch main journal size.

Since the deck height of both the 350 and 400 blocks are the same, you can use stock 5.565-inch 400 connecting rods since half the stroke increase (.135-inch)

is exactly the difference between the 5.7 and 5.565-inch rods. This will allow you to run a production style 4.030-inch piston to save the cost of a custom piston. Unfortunately, this produces a rather short 1.48:1 rod length to stroke ratio.

If your budget can afford a slightly more expensive piston, then you can step up to a number of custom pistons

available from JE, Ross or Speed-Pro that move the wrist pin up (a shorter compression height) that will allow you to add a longer 5.7-, 5.850- or even a 6-inch connecting rod. The longer connecting rod is generally considered an advantage, but rods longer than 5.7 inches means investing in aftermarket rods that cost more.

Advantages—Obviously, this engine combination offers tremendous opportunities to build it in a number of combinations depending upon the depth of your wallet. If you can afford such luxuries, there are a number of aftermarket companies offering upscale crankshafts from new cast cranks, to 4130 and 5140 steel cranks, up through Callies' Pro Street cranks that are an intermediate step between 5140 and the ultimate-strength 4340 steel crankshafts like those from Lunati or Crower.

Another feather in the 383's cap is the longer stroke typically improves low-speed torque both because of the longer arm and the added displacement. For street engines, this is a real benefit, since street engines spend a majority of their time at less than 3000 rpm. Combine this displacement with good aluminum heads, a hydraulic flat tappet cam of around 224 degrees of duration at .050-inch, and you've got a stout street engine. Torque comes in around 440 lbs-ft. while cranking around 425 horsepower.

406

Of course, if you're willing to run a 3.75-inch stroke crank, why not just add the larger 4.155-inch bore of a .030-over 400? Performance wives' tales have exaggerated the overheating problems associated with the 400's siamesed bores. This may have stemmed from hot rodders adding different heads and neglecting to drill the required steam holes between the center three cylinders. When a 400 block is properly machined and doesn't suffer from core-shifted thin cylinder walls, you can build an excellent torque thruster

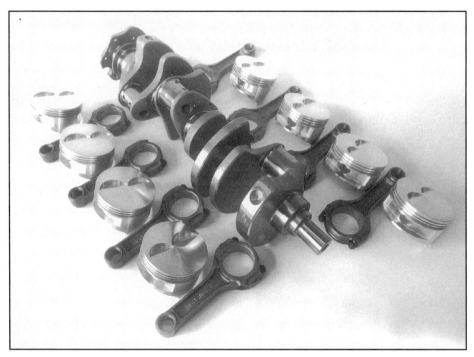

The displacement king of the production block Chevy is my 420-incher. Plug in a custom 3.875-inch stroke crank with custom 4340 steel rods and custom pistons and you have a street engine that can twist torque and horsepower numbers that will rival a big-block!

from a basic 400.

Cast nodular iron crankshafts can handle rpm well above 6000 rpm, but the 406's claim to fame is torque. Most street 400's tend to shine in the torque department with a peak torque around 4000 rpm, but unless fitted with excellent cylinder heads, horsepower usually does not follow through, especially with flat-tappet camshafts of less than 230 degrees at .050 inch. With mildly pocket ported iron 441 heads, it is possible to make 450 lbs-ft. of torque, while the horsepower comes in slightly above 400. Certainly more power is possible, and 500 lbs-ft. is attainable, although on pump gas it might be difficult.

420

Not too long ago, a 427 was the ultimate in large cubic-inch big-block displacement. Now, with help from a stroker crank, you can build a 420 cubic-inch small-block Chevy that weighs within a few pounds of any other small-block. Unlike the previous small-block bore/stroke combinations, this mega-Mouse requires a custom steel crankshaft with a stroke of 3.875 inches. A number of aftermarket crankshaft companies such as Callies, Crower and Lunati offer this stroke as a forging rather than a more expensive billet crank. I complement this stroke with a 5.850-inch connecting rod that results in a 1.51:1 R/L ratio that isn't ideal, but then this engine is usually intended as a torque monster rather than a high rpm horsepower engine.

Drawbacks—The stroke is limited by connecting rod bolt clearance problems with both the oil pan rail and the camshaft. Because of this, specially clearanced 4340 forged stroker rods are usually required to generate the necessary clearance. Custom pistons are also necessary to accommodate the long stroke and rod combination. As you can see, this hardly qualifies as an inexpensive engine, and the cost of just the crank, rods and pistons can exceed $2800. But comparing the weight savings of this "big" small-block versus a 427 Rat motor, there are reasons for going with the stealth factor of a big small-block.

Big Numbers—I have built a number of these engines using the ACCEL SuperRam electronic fuel injection intake I designed. This combination of cubic inches and intake manifold creates outstanding torque. One roller-cam 420 with ported Air Flow Research heads generated 407 lbs-ft. of torque at 1600 rpm. This engine also cranked out 542 lbs-ft. of torque at 4500 and 525 horsepower at 5500 rpm! Those are Rat motor numbers from an 11:1 small-block.

Rocket Block—Of course, if you're not interested in hiding the fact that you've invested your next two years' paycheck in an engine, you can up the

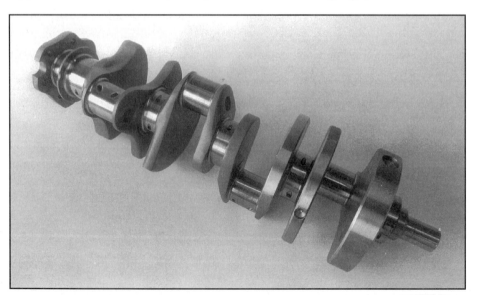

The beauty of the small-block is its amazing interchangeability. Grind down the 2.65-inch 400 main journals to the 350's 2.45-inch size and stuff it in the block for an instant 28 cubic inches to 383. Keep in mind that adding stroke and a larger bore will increase compression.

small-block ante with Oldsmobile's Rocket block casting. This block is based on the small-block Chevy but features a raised camshaft and clearanced oil pan rails that will allow you to build a ridiculously large small-block of up to 468 inches. We'll touch on a few details about the Rocket block in Chapter 3 on cylinder blocks.

INCHES ARE EVERYTHING

As you can see, there's plenty to choose from in different displacement small-blocks. While your imagination may run away with you, realism and the depth of your wallet will dictate what you can afford to build. It's no revelation that larger engines make more power. Typically, larger displacement engines generate great torque, but sometimes suffer in horsepower-per-cubic-inch because the longer strokes create additional piston speed that can eat horsepower at higher engine speeds.

It's also worth noting that there are not huge gains to be made with these displacement increases. It's better to view these increases as incremental steps rather than huge power increases. For example, the typical 383 can make 440 lbs-ft. of torque while the 406 will increase that to perhaps 460 to 480 lbs-ft. given the same equipment. One way to gauge the power your engine could make is to apply the "1.1" factor. A well-built street engine running a balanced package of components on pump gas will generate approximately 1.1 horsepower per cubic inch. The same tends to hold true for the torque. Divide out the typical numbers we've detailed in this chapter and you'll see these engines all hover around the 1.1 number. An outstanding engine will generate more than 1.1 while a weaker engine will fall a bit short of the mark. Overall, the 1.1 factor is a good yardstick for power compared to other small-blocks of differing displacements. ■

FACTORY DISPLACEMENT COMBOS

This chart will outline the lineage of the small-block Chevy ancestry from the first 265 in 1955 to the 305 in '76. There are actually a couple of variations in this chronology with different main journal sizes, but we'll stick with just the bore and stroke limitations on this chart. All dimensions are expressed in inches.

DISPLACEMENT	YEAR	BORE	STROKE
265	'55- '56	3.75	3.00
283	'57- '67	3.875	3.00
327	'62- '69	4.00	3.25
302	'67- '69	4.00	3.00
350	'67- present	4.00	3.48
307	'68- '73	3.875	3.25
400	'70- '80	4.125	3.75
262	'75- '76	3.671	3.10
305	'76- present	3.736	3.48

POPULAR PERFORMANCE DISPLACEMENTS

Hot rodders are always looking for cheap power. The best way to do this is by building a "stroker" engine, especially when you can do it with factory parts. Chevy did this in 1967 when they created the 302 by combining a 283 crank with a 327 block in order to meet the 305 cid requirements of SCCA Trans-Am racing. The following are some of the more popular bore and stroke combinations that hot rodders have created over the years. All dimensions are expressed in inches.

DISPLACEMENT	BORE	STROKE
355	4.030	3.48
372	4.125	3.48
377	4.155	3.48
383	4.030	3.75
388	4.060	3.75
406	4.155	3.75
415	4.155	3.825
420	4.155	3.875

3 CYLINDER BLOCKS

When it comes to building an internal combustion engine, you always start with the same foundation—the cylinder block. While it may appear that all small-block Chevy blocks are the same, that generalization couldn't be further from the truth. Currently, there are well over a dozen different major production-based bore/stroke and main bearing diameter small-block Chevy cylinder blocks. Then add the 1986 and later one-piece rear main seal blocks. Then there's also five different Bow Tie cast iron cylinder blocks, the three aluminum Bow Tie blocks, the four small-block-derived Oldsmobile Rocket Blocks and the aftermarket aluminum blocks, and you can see that there's plenty of variety in the small-block Chevy arena.

It would probably take a book this size just to classify all the different small-block Chevy cylinder blocks and the details that make them special. For sanity's sake, we'll concentrate mainly on the production 350 and 400 blocks and take a glance at a few of the Bow Tie cases as well as the wild Rocket Blocks just to outline what's available. We'll focus mainly on the features and functions of the various blocks. Machining operations and special modifications will be left to later chapters.

The wonderful thing about the small-block Chevy is that even with the

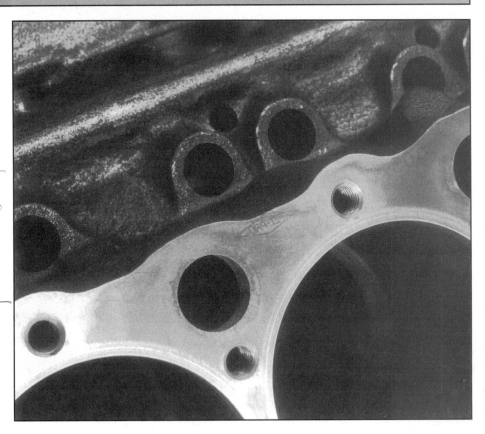

The cylinder block is the foundation that you build upon. Over 40 years of production has produced a wide variety of blocks to choose from. We've narrowed them down a bit to help make the decision easier. Don't skimp here. Get the best block you can afford.

tremendous variation in cylinder blocks, many components like camshafts, cylinder heads, valvetrain, and many more pieces are completely interchangeable between these blocks. The small-block is also interchangeable externally, with virtually all of its variations. The one exception to this is the 1955 through 1957 265 and 283 engines that are not fitted with the traditional side engine mounts. Interchangeability is the single most basic

factor of the small-block that has made this family of engines so popular with hot rodders.

SMALL JOURNAL BLOCKS

In the beginning, all small-blocks were created equal. That means that they all were equipped with main bearing inside diameters of 2.300 inches. As you can see by the nearby chart, this was the way all small-blocks were constructed through

The only way to ensure the block you wish to use is the correct bore is by measuring the bore diameter with a dial bore gauge. Measure the bore diameter at the top of the ring travel. This is the point of maximum bore taper and will dictate the amount of overbore required to ensure a round cylinder. Be forewarned however, worn cylinders often require more of an overbore than indicated by the dial bore gauge.

BLOCK SPECS

DESCRIPTION (Bare block)	DIMENSION (Inches)
Length:	21-3/4
Width: (at bellhousing)	16-3/4
Height: (w/o main caps)	9.80
Bore Spacing:	4.4
Deck Height:	9.025
Weight:	152*

*Approximate weight. Will vary depending upon year of casting, displacement and other variables. Iron Bow Tie blocks weigh approximately 182 pounds.

JOURNAL DESCRIPTION	BEARING I.D. (Inches)
Small Journal (1955 - '67)	2.30
Medium Journal (1968 - current)	2.45
Large Journal (1970 - '80)	2.60

1967. This included the 265, 283, the '67 302, and all 327's cast before 1968. These blocks are not nearly as desirable today because of their small displacement and also because over 30 years of hot rodding have made them difficult to find. It's important to be able to identify a small journal block if just to know what to avoid or to be able to pick one out of a pile should the need arise.

Identification

The small-journal 327 block, which was shared by the '67 Z/28 302 engine, offers a respectable 4.00-inch standard bore although it would be next to impossible to find a 327 block now that is anywhere near 4.00 inches. The early blocks were all fitted with a number of casting differences that make them easy to identify. All small-journal blocks were equipped with front and rear crankcase breather holes. The rear hole was equipped with a road draft tube that dumped crankcase vapors overboard. Engines beginning in '66 in California were equipped with tubes that directed these vapors through a positive crankcase ventilation (PCV) valve and back into the air cleaner. All of these blocks were 2-bolt main versions. Factory four-bolt main blocks did not arrive until the medium journal blocks after 1967. These blocks can be converted to four-bolt mains with aftermarket steel main caps although for street use this isn't necessary.

These blocks of the Fifties and Sixties also enjoyed the strength of thicker cylinder walls and better cast iron which

Pre-'68 blocks can be quickly identified by both the road draft tube fixture in the rear of the block next to the distributor hole and the oil fill hole in the front of the block. This identifies this block as a small main bearing journal block.

The biggest change to the block occurred with the introduction of the one-piece rear main seal design in 1986 (top). This necessitated a block casting change, a new rear seal, different crankshaft and oil pan. None of these components will interchange with the older 2-piece rear main seal blocks. However, you can use the older 2-piece crankshaft and oil pan with the '86 and later blocks with an adapter.

made them capable of large overbores. Many a 283 has been bored .125-inch oversize. The later 327 blocks are capable of .060 inch, but more than that requires a sonic check for cylinder wall thickness on the thrust side of the cylinder, which is the inboard wall on the left (driver's) bank and the outboard wall on the right (passenger's) bank of the block.

We've described these small-journal blocks mainly for identification purposes. Frankly, these early blocks are increasingly harder to find, becoming more expensive, and limit engine size since there is no logical reason to add a stroker crank to one of these early blocks. That's not to say that you shouldn't build one of these engines if your cousin leaves a grungy block on your doorstep. It's just that these early engines don't offer as many advantages or opportunities as the later small-block castings.

MEDIUM JOURNAL BLOCKS

Without question, the medium journal small-block, mainly the 350, is the most popular small-block ever built. Beginning in 1968, Chevrolet changed the main journal diameter of its crankshafts to 2.45 inches in an effort to strengthen the

longer stroke crankshaft offered with the 350 engine. This production line change also affected the 1968 and later 302 and 327 as well, which is why you should be careful when mixing 327 engine components, since the crankshaft main and connecting rod journal changes do not interchange.

010 Casting—There have been literally dozens of different castings of the medium journal cylinder block, but the variations are limited to a few significant changes. Other than the switch to larger journals, the most significant modification was the addition of 4-bolt main caps beginning with high performance 350 passenger car engines and the truck applications in '68. Many hot rodders believe that all "010" blocks (the last three digits of the block's casting number) are equipped with 4-bolt mains, but this has been proven to be untrue. Perhaps a more accurate statement is that many of these blocks are fitted with 4-bolt mains. Unless you pull the pan, don't be suckered into buying a 4-bolt block strictly on the 010 myth.

Two-Bolt vs. Four-Bolt

As with the small-journal blocks, there is nothing wrong with a 2-bolt main

block for all but the most exotic street small-block. However, if a 4-bolt block is on your list of "must-have" items, there are a number of options. Chevrolet offers nodular iron four-bolt main caps that can be adapted to your two-bolt block. This cap offers parallel main cap bolts and is a semi-finished cap which means that the block will have to align bored and the ends of the cap must be machined to properly fit the block. A number of aftermarket companies, such as Summers Brothers and others, offer 8620 steel main caps in either parallel bolt or splayed bolt configurations.

Chevrolet also offers a 1010 steel splayed, four-bolt cap for the medium journal block. Splayed bolt caps are inherently stronger than parallel bolt caps since the angled outboard bolts anchor

A quick way to identify a 400 block is by its wider, big-block looking main caps whether 2-bolt or 4-bolt. The block on the left is the 400 casting while the block on the right is a 350 block.

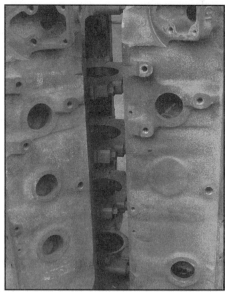

Don't rely on the wives' tale that all 400 blocks have three freeze plugs. There were many 400 blocks cast with only 2 freeze plugs.

into the thicker area of the cylinder block. Regardless of which cap design you choose, changing main caps is best left to a qualified machine shop. Lingenfelter Performance Engineering offers this conversion on any small-block.

Best Choices

The most popular medium journal engine is easily the 350 with its 4.00-inch bore. As evidenced by the bore and stroke chart in Chapter 4, there are also four smaller factory bore diameters in the medium journal block family. Of these, the 305 with its 3.736-inch bore and the 307's 3.875-inch bore are the most common. Either of these engines make great street small-blocks although they do suffer the same small displacement disadvantage. If you are looking for an inexpensive small-block and ultimate power is not a priority, either of these engines can probably be picked up cheap since they are not high priority engines within the hot rod community. Keep in mind, however, that parts and machine

work cost just as much (if not more) to build a 307 as it does to build a 350. The choice is yours but often saving money with a small displacement engine doesn't pay off when it comes time to make power.

With the tremendous number of used blocks available, it should be easy to locate an inexpensive used foundation for your next high performance street engine. But if you want to start with a brand new block, Chevrolet does offer two medium journal cast iron blocks with either 2- or 4-bolt main caps in the new 1-piece rear main seal style block. There's also a 4-bolt main cap block in a 2-piece case through any Chevy dealer. Complete engine assemblies are also available.

LARGE JOURNAL BLOCKS

In 1970, Chevrolet introduced a third variation in the Mouse motor lineage. The 400 small-block became the largest displacement factory small-block and was built from 1970 until 1980. Pumping

the Mouse motor up to this displacement required significant internal changes while maintaining the block's external dimensions. The internal changes demanded a larger main journal of 2.65 inches to support the load of the longer 3.75-inch crankshaft stroke. The big 4.125-inch standard bore also demanded that the bores be siamesed, which means there are no water jackets between the adjacent cylinders. Some hot rodders feel this is a problem and avoid the 400 block, using veiled references to scored cylinder walls and chronic overheating as evidence. To the contrary, the siamesed cylinder walls tend to improve cylinder integrity and cooling the big 400 is not a problem with some simple modifications to the cylinder heads.

Identification

Identifying the 400 block is relatively easy. Some enthusiasts claim you can spot a 400 by its three freeze plugs, which is true. However, there are also a few 400 blocks that were cast with only two freeze plugs on each side of the block. The easiest way to spot a 400 block with the heads removed is by the characteristic siamesed cylinder walls,

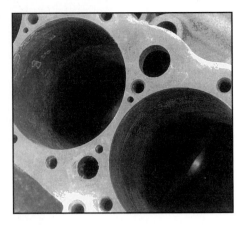

The 400 block is equipped with additional "steam" holes drilled into the deck surface to vent potential steam pockets into the heads from the siamesed cylinder bores. If you switch cylinder heads on a 400, make sure the head gasket and head have matching steam pocket holes.

evidenced by the reduced space between adjacent cylinders. If just the pan is removed, all 400's have a notch at the outboard bottom of each cylinder bore used for additional connecting rod clearance for the longer stroke. Of course, if you have a dial bore gauge, the 4.125-inch standard bore is the best indicator.

Coolant Holes—Additional coolant or steam holes in the 400 block's deck surface also differentiate the 400 from its smaller main journal cousins. These steam holes are used to ventilate steam pockets out of the block into the cylinder heads. Overheating problems can be caused by using non-400 cylinder heads not fitted with matching steam relief holes. This may be one reason why the 400 has earned its undeserved reputation as prone to overheating. This is especially prevalent in low rpm street engines operating a majority of time below 3500 rpm. For higher rpm applications, I plug the large 3/4-inch coolant circulation holes located above the cylinder bores and redrill them with a smaller 3/16-inch hole sized to match the head gasket. This modification improves deck surface strength and helps prevent cracks between these holes and adjacent head bolts.

Some early 400 blocks were cast

TOO-COOL 400'S

For many years, the 400 has suffered from a reputation as an engine that tends to run hot in street operation. The "steam holes" found in between the siamesed bores of the 400 block serve to vent steam pockets that may form in low speed operation such as on the street. Higher rpm operation tends to sweep these steam pockets back into the coolant. However, swapping heads can often lead to overheating if these steam holes are not matched between the block, head gasket and the cylinder heads. Most aftermarket heads do not come with this steam pocket hole drilled into the water jacket through the deck surface. To drill this steam relief hole, use a 400 cid head gasket as a template and match the hole size.

Some 400 blocks also come with a coolant hole located between the two center cylinders to cool the two adjacent exhaust ports that create additional heat. The engineers at Fel-Pro have developed a solution to solve this overheating problem with some 400 engines. The procedure is designed to place more coolant around the adjacent exhaust ports to prevent the creation of steam that hurts the cooling process. If your 400 block is not equipped with this hole, use the drawing as a guide to position the hole. This will require drilling the 3/8 to 7/16-inch hole in the block, head gasket and the cylinder head to promote additional cooling. Use a hole punch or an empty bullet casing to punch holes in the head gasket.

One way to improve cooling efficiency in a 400 is to block off the large upper three coolant holes in the deck surface with 3/4-inch pipe plugs. Tapping this thin deck surface may be difficult, but will strengthen the block deck surface. Then drill 3/16-inch coolant passage restrictions in these 3/4-inch plugs to allow some coolant to pass through to the cylinder heads. These large 3/4-inch plugs can be left sticking slightly above the deck surface to be milled flush when the block surface is milled.

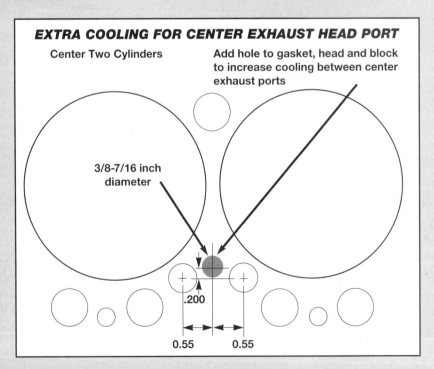

EXTRA COOLING FOR CENTER EXHAUST HEAD PORT

Center Two Cylinders

Add hole to gasket, head and block to increase cooling between center exhaust ports

3/8-7/16 inch diameter

.200

0.55   0.55

The shaded area indicates the hole that should be added to a 400 block located between the two center cylinders. The dimensions are indicated on the drawing. Some 400 blocks may already be fitted with this coolant passage. Make sure that the cylinder head and the head gaskets also match this coolant passage.

Production '87 and later small-blocks are also fitted with taller hydraulic roller lifters which require taller lifter bore castings. The lifter bore diameter is the same and older flat tappets can be used in place of the rollers, although there's little reason to do so. The blocks also feature specific machined bosses to locate the lifter retainer plate. Factory hydraulic roller lifters cannot be used in pre-'87 blocks due to their increased height.

without an additional coolant hole between the two center cylinders that was added on later model 400 blocks. This 3/8 to 7/16-inch hole is used to supply additional coolant between the center two exhaust ports. When building a healthy 400, this hole should be added to the block, head gasket and the cylinder head if the holes are not already present. This reduces the formation of steam pockets between the siamesed exhaust ports. This reduces the possibility that these two center cylinders will detonate.

While Chevrolet added additional material to the cylinder walls on the major thrust surfaces on the 400, it is common practice to avoid bores of more than 4.155 for a 400 intended for serious high performance. While there are plenty of examples of .060-over 400's, it's not recommended because thin cylinder walls tend to move around more under load, creating a more difficult sealing job for the rings.

Differences—The differences with the 400 continue with its four-bolt main caps. The 400's four-bolt main caps

cannot be interchanged with the two smaller journal engines since the larger 2.65-inch main journal diameter required the bolts to be spaced further apart. As with all production engines, only parallel four-bolt caps are utilized in the 400. High performance splayed caps can be added to the 400. If this procedure is to be performed, I recommend using the stronger 8620 steel alloy caps. The original outboard bolt holes must be plugged before the outer splayed bolts can be machined. This procedure, along with the parts, is not cheap. This should only be considered if the engine is to be subjected to competition or extreme high rpm use. Given the machining costs involved, it might be smarter to step up to a Bow Tie block with its inherent strength advantages rather than compromise with a production 400 block.

If serious performance is planned with a production 400 block, I recommend the 2-bolt main block. If for no other reason, the 2-bolts make it easier to machine the two outer bolts. The factory 4-bolt blocks also sacrificed some material around the

main webs that make the 2-bolt blocks actually stronger. It has been my experience that many knowledgeable circle track racers all prefer the 2-bolt 400 blocks.

1986 & LATER BLOCKS

The final major facelift performed on the small-block Chevy occurred in 1986 when Chevrolet changed to a one piece rear main seal design to reduce oil leaks. This design also necessitated a change to the crankshaft, which will be detailed in Chapter 4. In addition, Chevrolet also converted all of their passenger car cylinder blocks to roller lifters. This change included small machined bosses in the lifter valley to mount the roller tappet locator along with taller lifter bosses to accommodate the taller roller lifter body.

LT1

Falling into this late model one piece rear main seal category block is the 1992 and later LT1 small-block. While few of these engines have found their way into the hot rod world as yet, they do represent another departure from the typical small-block mainstream. The main redesign is a function of the way the cooling system is routed through the block and heads. In the traditional small-block, water is routed through the block and then the heads before it is returned to the radiator. The LT1 uses a reversed cooling system that routes the coolant first through the heads (where significant heat is located around the combustion chamber and exhaust ports) and then routed through the block before returning to the radiator. There are a number of other significant changes to the new LT1 such as a camshaft-driven water pump, front-mounted distributor and other modifications too numerous to mention here. Suffice to say that the LT1 is different enough that it should be considered an off-shoot of the small-block rather than a pure blood

ONE PIECE OR TWO?

PART NUMBER	DESCRIPTION
10108477	Stock Chevrolet aluminum retainer that mounts the one-piece seal in an '86 and later block using a one-piece seal oil pan. Also needed is the gasket (PN 12337823) and the one-piece seal (PN 0088158).
10051118	Chevy adapter that allows use of two-piece crankshaft in a one-piece seal cylinder block with a one-piece main seal oil pan. You will also need a 2-piece rear main seal, adapter gasket (PN 12337823) and a one-piece pan gasket (PN 14088505)
38315 (Moroso)	Allows use of two-piece seal crankshaft in a one-piece seal block with a two-piece rear main seal oil pan.
02-5702 (Diamond Racing)	Same adapter as above

descendant.

All of the non-LT1 one-piece rear main seal blocks will still accept most of the typical small-block components. The earlier crankshafts and oil pans will also interchange, but you will need a different GM rear main seal adapter to allow the use of the earlier components. For those contemplating using an '86 or later block, remember that the smaller diameter crankshaft flange will require a later flexplate/flywheel and the oil pan may not clear the crossmember of early Chevy musclecars. Other than these two major changes, the '86 and later engines will directly interchange in terms of size, bellhousing bolt pattern, and engine mount locations. If you desire a brand-new casting, Chevrolet offers both two- and four-bolt main cap blocks with the one-piece rear main seal.

The rear main seal adapter question can get a bit complicated. Chevrolet produces an aluminum adapter to mount the stock one-piece rear main seal to the '86 and later blocks. They also offer an aluminum adapter that configures the block to use the one piece oil pan with the older two-piece seal crankshaft. Companies such as Moroso, Diamond Racing and Katech offer an adapter that will mount the older two-piece seal crankshaft and oil pan to the late model one-piece block. To say that small-block Chevy interchangeability is becoming more confusing might be an understatement.

BOW TIE BLOCKS

In addition to production-based cylinder blocks, the power demands placed on production blocks demanded the creation of Chevrolet's Bow Tie block. The major Bow Tie block improvements over stock involve the use of high nickel content cast-iron, thicker cylinder walls and deck surfaces, as well as a number of other changes that make the difference between compromising with a production block and using the casting that championship drag racers and NASCAR stock cars win with every weekend. How strong are these Bow Tie blocks? I've been able to coax over 1400 horsepower from a Bow Tie block with no durability problems.

I won't cover all the details since the Bow Tie seems to be in a constant state of change. Currently, there are five different cast iron and now four Bow Tie aluminum cylinder blocks.

Iron Bow Ties

The iron blocks offer numerous variables, including two- or four-bolt

Chevy's Bow Tie block can be quickly identified by its cast-in Bow Tie in the right hand, passenger side of the block. These new second generation blocks offer thicker cylinder thrust surfaces (.340-inch nominal), thicker deck surfaces, and one piece rear main seals. Exceptions to this are the "Qualified" race blocks designed with two-piece rear main seals and dry sump oiling.

There are three different Bow Tie blocks available with splayed, 4-bolt main caps made of 8620 alloy steel. For a street engine, these killer NASCAR oriented main caps aren't necessary but do offer ultimate reliability compared to the nodular iron main caps. Note the "splayed" bolt arrangement. This happens to be a dry sump block, identified by its solid rear main cap.

main caps, splayed main caps, bore size, siamesed cylinder walls and whether the block is intended for wet or dry sump oiling! Instead of a long, drawn-out explanation, I'll let the Bow Tie block chart on p. 26 explain it all. The latest version race-prepared Bow-Tie blocks offer priority main oiling where the oil is routed directly from the center oil gallery to the main bearings via two intersecting 5/8-inch passages. This eliminates the annular grooves in the cam bearing bores that restricts oil to the mains in production engines.

In addition to these new variations, the race-prepped Bow Ties offer a camshaft thrust flange machined for a thrust retainer that will offer improved control over cam walk. These blocks also guarantee at least a .175-inch wall thickness at the thrust surface at the maximum 4.160-inch bore. Additional material is added around the front and rear cam bearing bores. Add to this larger

lifter bosses and a provision for a rear camshaft block-off plate and it's clear the new Bow Ties are dressed for success.

Aluminum Bow Ties

If you're obsessed with maximizing power-to-weight, Chevy offers several aluminum Bow Tie blocks that can shave over 90 pounds off their cast-iron cousin and dramatically lighten your wallet at the same time! Chevy offers four different blocks, but only one is intended for wet sump use. It is equipped with splayed four-bolt main caps and a 2-piece rear main seal. This block is race-prepped machined although certain modifications such as drilling for a dip stick will still be needed for street use. While the 90-pound weight loss is great, I don't think it is a good tradeoff for the money, because power suffers from cylinder wall instability and major heat rejection at maximum power levels. Chevy has also just introduced a 9.525-inch tall deck,

raised cam, aluminum block with widened pan rails that offers even larger displacement potential for those crazed few looking for the ultimate in a flyweight, large-by-huge Mouse motor.

THE ROCKET BLOCK

If the array of Chevrolet Bow Tie blocks isn't enough to make your head spin, then you're probably not ready for the Rocket Block option. These babies should be considered the ultimate in cast iron cylinder blocks and are intended for professional competition. Nevertheless, the Rocket Block offers the opportunity for obsessive-compulsive power mongers to conjure up a 468 cid small-block based on its tall 9.325-inch deck height. There are actually four different Rocket blocks, two dry sump and two wet sump models. Unfortunately, the Rockets weigh in at a hefty 200 pounds due to all that added iron. The Rocket Block is designed to use a small-block Chevy crankshaft, rods,

If you're looking for the ultimate in iron cylinder blocks, the Olds Rocket block may be your ticket. The block is capable of a large bore and stroke but is also more expensive and heavier than the Bow Tie block. In most cases, the Bow Tie is sufficient.

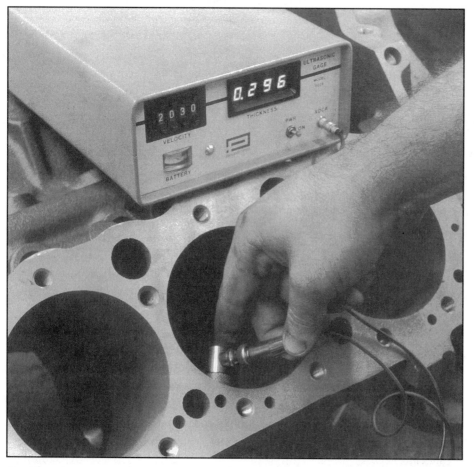

If a block is questionable, sonic checking is one way to ensure reliability. I prefer .200- to .240-inch of cylinder wall thickness on the major thrust side on production 350 blocks. The fore and aft, non-thrust walls are thinner at .090-.110-inch. Bow Tie blocks are thicker although Chevy recommends not boring more than 4.160-inch. The major thrust surfaces, where the piston pushes against the cylinder wall, are the inboard wall on the left (driver's) bank and the outboard cylinder wall on the right cylinder bank.

cylinder heads and camshaft while allowing more latitude in creating many more displacement options. Be forewarned however, these Rockets live up to their name with an astronomical price tag!

SELECTING A USED BLOCK

The classic hot rod approach is to create a masterpiece of power from someone else's discarded junk. Used small-block Chevy blocks are perhaps the easiest engine on earth to find since there isn't a bone yard or recycling yard that doesn't have at least a few hundred lying around. Thus it pays to know what you're looking for when it comes time to locate a core suitable for your high performance needs. A little time spent searching for a prime candidate almost always pays off later in a solid performance foundation.

Whether the engine is still partially assembled or merely a bare block, there

Pressure checking is an excellent way to avoid spending hundreds of dollars of machine work on a block Swiss cheesed with porosity. I use this fixture to pressurize the cooling system to check for leaks. If the block leaks, don't use it.

The center, main oil gallery feeds oil directly from the oil filter to the front of the block. The Number One front main bearing and first two rods feed off this passage drilled up from the main bearing web. This intersecting oil gallery can be partially blocked if a too-long pipe plug is used to seal the main oil gallery. Measure the depth of the plug versus the tapped hole.

Part of block preparation includes grinding all casting flash from the oil return holes in the lifter valley to prevent cast iron and cast iron foundry sand from finding its way into the oiling system and damaging the bearings.

Many small-blocks are drilled for only the straight across style starter motor bolt pattern. This only allows the use of the small 153 tooth flywheel. In order to use the larger 168 tooth flywheel (needed for the larger 11-inch clutch), the block must be equipped with the 3-bolt pattern that allows the use of the offset starter motor bolt pattern.

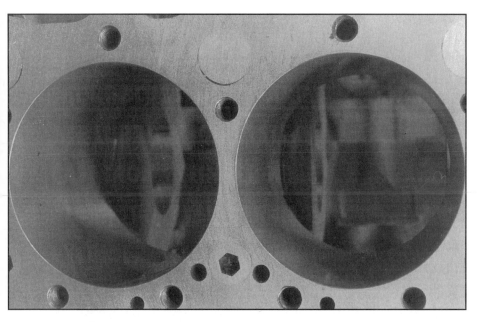

Moroso and others offers 3/4-inch pipe plugs that can be used to block off the top row of large coolant holes in the cylinder block deck surface. These holes can then be redrilled with smaller holes that match the restrictions in the head gasket. These large pipe plugs also improve block deck surface strength.

are a number of key points to look for when looking for a strong block. Surface conditions like excessive internal or external rust is a good place to start, along with a quick inspection of the main caps to ensure the block has not been damaged by a spun bearing or lack of oil. Blackened or missing main caps are red flags that should tell you to look elsewhere. Less obvious but also dangerous are main caps that don't fit snugly in the main cap saddle.

Assume with any used engine that it will require an overbore of no more than .030-inch. If the engine has a serious ledge just below the top of the cylinder, this usually indicates serious cylinder bore wear that might require a larger overbore. Especially with thin-wall, late-model engines cast after the mid-70s, this

might be a mistake. Also inspect the block for cracks between deck coolant and the head bolt holes. Cracks also tend to occur at the base of the cylinder bore adjacent to the main bearing webs. Often these holes will be masked by oil or sludge until the block is thoroughly cleaned.

Core shift is another area to investigate. Good blocks are easy to spot by looking for a concentricity in the machined cam bore area in the front of the block. While hot tanking a block makes it much nicer to work on, it also presents another opportunity to search for flaws before expensive machining begins. If the block is to be used in a moderate to high horsepower application, you should invest in a cylinder wall sonic check and perhaps a pressure check that will indicate a hairline fracture near a water jacket. There's nothing more frustrating than investing good money in machine work only to discover a cracked block that relegates it to the scrap pile.

Investing the extra time in choosing the correct cylinder block will never be the most glamorous task, but like the house built on a strong foundation, a sound cylinder block will withstand the rigors of the most powerful small-block. ∎

CAST IRON BOW TIE BLOCKS

Part Number:	110051181	10051183	10185047	24502501*	24502503*	24502525*
Cyl. Wall:	Non-Siamesed	Siamesed	Siamesed	Non-Siamesed	Siamesed	Siamesed
Deck Height:	9.025	9.025	9.025	9.025	9.025	9.150
Cyl. Bore:	3.725-4.020	4.000-4.160	3.995-4.120	3.750-4.020	3.995-4.120	3.995-4.160
Main Cap Bolts:	2	2	4	4	4	4
Main Cap Style:	Straight	Straight	Straight	Splayed	Splayed	Splayed
Cap Material:	Cast Iron	Cast Iron	Nodular Iron	8620 Steel	8620 Steel	8620 Steel
Main Journal:	2.45	2.45	2.45	2.45	2.45	2.45
Oil Sump:	Wet	Wet	Wet	Wet	Wet	Wet
Rear Main Seal:	1 or 2 pc	1 or 2 pc	1 or 2 pc	2 pc	2 pc	2 pc
Weight:	181	181	182	180	187	191
Features:				CNC Machined	CNC Machined	CNC Machined

* The last three Bow Tie blocks are Race-Prepped blocks, offering computer numerical control (CNC) finish machine work completed at the factory. These blocks offer priority main oiling, .360-inch taller lifter bores, fully machined pan rails, bellhousing flange and front face among other changes. The block still requires finish boring and honing. According to Motorsports Technology Group Block Engineer Larry Kubes, these blocks can save the average enthusiast $500 or more in finish machine work costs over the basic Bow Tie block, work such as installation of the 8620 steel splayed main caps and finish line honing that's no longer necessary.

PRODUCTION CYLINDER BLOCKS

Part Number:	10066034	10066098	10105123
Material:	Cast Iron	Cast Iron	Cast Iron
Bore Diameter:	4.00"	3.74"	4.00"
Cyl Wall:	Non-Siamesed	Non-Siamesed	Non-Siamesed
Main Caps:	Cast Iron	Cast Iron	Cast Iron
Main Bolts:	4	2	4
Crank Journal Dia.:	2.45"	2.45"	2.45"
Rear Main Seal:	2 piece	1 piece	1 piece

CRANKSHAFTS AND BEARINGS 4

If you were to line up a dozen small-block Chevy crankshafts, they might all look the same, but when it comes down to the important details, this is anything but a one-size-fits-all world. Between the questions of iron vs. steel, journal diameters, stroke lengths and internal vs. external balance, there is a world of difference and more than a little confusion. While there is probably more to know than you care to learn, if you build high performance small-blocks, these are questions that need answers. This chapter will dispel many of the myths and present solid information to determine which crankshaft is best for your engine.

The good news is that even the lowly production Chevy iron cranks are more than adequate for high performance street use. While those romantic steel aftermarket crankshafts are appealing, the best advice for the small-block builder constructing a typical small-block making even over 400 horsepower is that stock cranks are more than up to the task. So why bother learning all the details about high performance crankshafts? Think of this chapter as inexpensive insurance against making the wrong decision. Taking the time to become familiar with crankshaft details now can save you untold grief later.

IRON VS. STEEL

Small-block crankshafts are made of two different materials—iron and steel.

Small-block cranks all look the same at first, especially in photos. But there are dozens of small yet significant differences that determine which crank is best for your application.

The two metals are vastly different. From the first 265 to roll out of the engine plant until about 1963, all production Chevy cranks were forged steel. Chevy soon learned that nodular iron was a suitable substitute that could be cast rather than forged, at a reduced cost. All production small-block Chevy cranks built today are nodular iron. Steel cranks offer dramatically higher tensile strength and tend not to be as flexible, which is an advantage in high output racing

applications. Just to add to the confusion, it is this nodular iron flexibility that allows it to be used in street high performance applications with great success. But when it comes to high output race engines, a floppy crankshaft is hardly preferable.

Among both Chevrolet and aftermarket steel crankshafts, there are many variations of material. All production small-block Chevy cranks are forged from 1053 steel. If these steel alloy

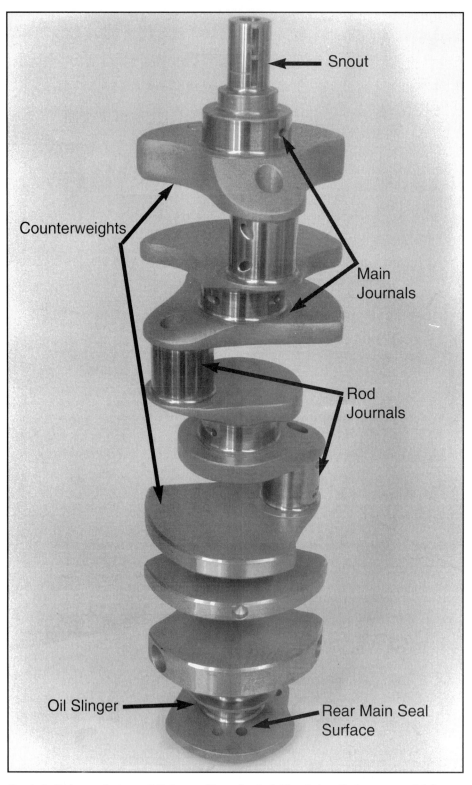

Snout

Counterweights

Main Journals

Rod Journals

Oil Slinger

Rear Main Seal Surface

Crankshafts have a lexicon all their own. The call-outs in the photo will give you a quick lesson in crankshaft vocabulary. A working knowledge of crankshaft definitions makes it easier to work with machine shops and aftermarket crankshaft companies.

when it comes time to purchase a high performance steel crank.

Heat Treating

Heat treating plays a critical role in crankshaft strength. All production Chevrolet crankshafts are heat treated to improve surface hardness, usually with some type of thin, surface treatment called nitriding. Usually when a crankshaft is machined to true the bearing journals, this heat treatment is removed since it is usually not more than .005-inch thick. Used crankshafts can be heat-treated using this same nitriding process but for a mild street engine this isn't necessary. Thousands of small-blocks are rebuilt every year without heat treating.

However, many crankshaft manufacturers and metallurgists suggest that the type of heat treatment can mean as much to the ultimate strength of a crankshaft as the base alloy material. The following is a quick look at a few of the more popular heat treatments and their advantages or disadvantages.

Hard Chroming—This process applies a relatively thick finish that produces a very hard and durable surface to the crankshaft journals. This prevents minor damage to the journals, but also reduces ductility. Hard chroming also makes repair difficult, if not impossible. This process is not recommended for either street or race cranks.

Nitriding—A heat treating process that produces a relatively thin (.006-.010-inch) but harder bearing surface. This process is perhaps the most popular case hardening procedure. Machining the journals usually removes this thin case hardening.

Deep Case Nitriding—Similar to nitriding except that more heat is used creating a thicker case hardening. This also creates a less ductile crankshaft that is difficult to repair.

Tufftriding—This procedure is an older case hardening process using a salt bath procedure that can cause chemical

number designations are confusing, we've included a chart on the next page that lists the more popular steel alloys in terms of their relative strengths. It's not important to know the makeup of each individual alloy but it is important to know that a 4340 steel crank is inherently stronger (and more expensive) than a 5140 steel forging. If you understand the relative rankings of these steel alloys, it will help

IT'S ALL IN THE ALLOY

There are literally dozens of steel alloys as identified by the Society of Automotive Engineers (SAE). Fortunately, there are only a few that are popular alloys useful for a crankshaft. Rather than attempt to break down each of these alloys, we'll list the popular alloys by their relative strengths. Price is directly proportional to the strength. In other words, strength costs money—how strong is strong enough?

ALLOY	DESCRIPTION
1010	The basic alloy steel. This is the alloy used in production Chevrolet crankshafts.
1053	The next step up in higher tensile strength. This is the material used in Chevrolet's over-the-counter steel cranks.
530A36	This is an English alloy designation used by Callies that balances strength and price.
5140	This alloy is one step below the optimal 4340 offering a great compromise between strength and cost.
4340	This is the ultimate high strength alloy. Higher strength alloys are more brittle. The ultimate crank is a billet 4340 crank.

The first major difference in crankshafts is cast versus forged. A cast crankshaft can be quickly identified by its thin, narrow casting line, as evidenced by the crank on the left. The crank on the right is an example of a forged crank with its wide forging mark. Forged cranks also tend to ring when tapped lightly with a hammer. Cast cranks react more with a thud. Aftermarket steel cranks often don't reveal a wide forging mark since it is removed during machining.

etching if the salt is not thoroughly removed. For this and environmental reasons, Tufftriding is no longer the case hardening process of choice.

PRODUCTION CHEVY CRANKS

There are literally dozens of different small-block Chevy crankshafts. As we mentioned, early Chevy cranks were all forged steel, but by 1968 the majority of production cranks were nodular iron. The main differences between small-block crankshafts are their stroke, main and rod journal diameters. I have placed these in an easy-to-read chart on p. 31. The significant differences besides stroke are the main and rod journal diameters and the 1986-and-later one-piece rear main seal crankshafts for the 305 and 350 engines.

Since fewer street engines today are built smaller than 350 cubic inches, I won't detail the earlier 283, 302, and 327 crankshafts. There are some subtle differences within the later cast-iron crankshafts, including 305 versus 350 crankshafts. Both of these cranks offer the same 3.48-inch stroke and 3932442 casting number. However, the 305 cranks are balanced for a lighter piston and rod assembly. For mild street use, you can use the 305 iron crank in a 350 with good success. However, adding stronger rods and/or forged pistons can create balancing problems since the 305 crank's counterweights are of insufficient mass to allow easy balancing. The sidebar nearby reveals how to quickly distinguish a 305 from a 350 crankshaft.

Rear Main Seal Cranks

Besides this subtlety, the major change in small-block Chevy crankshafts since 1970 is the 400-style crank, which I'll address later, and the one-piece rear main seal crankshaft change in 1986. Chevrolet made a major change to its crankshaft design with the creation of the one-piece rear main seal, designed to reduce the

As illustrated in the crank chart, the small-block Chevy crankshaft has three different main journal diameters—small (2.30-inch), medium (2.45-inch) and large (2.65-inch). The best way to determine the crank you have is to measure the crank's main journal diameter with a micrometer. Measuring each journal will indicate whether the crank should be turned or merely polished before using.

After a crank is machined, it's preferable to radius the oil holes in both the main and rod journals to make it easier for oil to enter the tight clearance area between the bearing and the crank journal.

THE IMPOSTOR

One of the subtle differences that lie in wait for the small-block Chevy engine builder is the great 305 vs 350 crank caper. Both engines use crankshafts with the same main and rod journals, stroke and even the same casting number (3932442). Unfortunately, even though they can and are used interchangeably, the crankshafts are not identical. The difference lies in the way the cranks are balanced for the different reciprocating masses of the two engines. The main difference between the 305 and 350 is the diameter and weight of the 350's larger bore pistons. This crank is also used in the obscure '79-'80 267 cid 3.48-inch stroke engines.

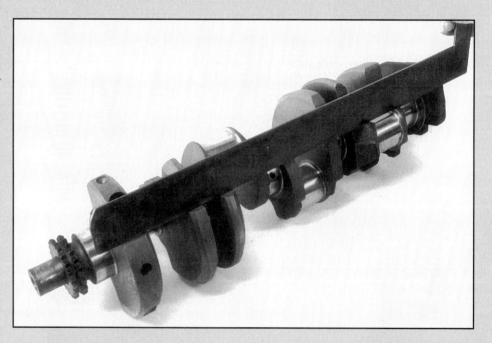

In mild street applications, the 305 crank can be used with stock weight replacement pistons allowing the rotating mass to be properly balanced. Unfortunately, when high performance rods and/or pistons are added, the weight difference will require expensive "heavy metal" to properly balance the crank. A 305 crank is not the best choice in this case. The best procedure is to identify a 305 crank before the expensive machine work is completed.

The best way to spot a 305 without having to spin it up on a balancing machine is to lay the crank down horizontally and position a straightedge along the machined surfaces of the crank as shown. A 305 crank will allow the straightedge to lay flat across all five points. A 350 crank will not allow the straightedge to sit flat.

CRANK CALLS

Small-block Chevy cranks can be differentiated into three basic categories of small, medium and large journals. These sizes refer to the main journal diameter. Within these three groups are a number of different stroke lengths. Rod journal diameters also vary, although the medium and larger main journal cranks share the same rod journal diameter. A recent addition to this group are the '86 and later one-piece rear main seal crankshafts. These one-piece seal cranks are listed separately since they do not interchange with earlier cylinder blocks.

SMALL JOURNAL CRANKS

Engine	Main Journal Diameter	Rod Journal Diameter	Stroke
283 ('57-'67)	2.300	2.00	3.00
302 ('67 only)	2.300	2.00	3.00
327 ('62-'67)	2.300	2.00	3.25

MEDIUM JOURNAL CRANKS

Engine	Main Journal Diameter	Rod Journal Diameter	Stroke
262 ('75-'76)	2.45	2.100	3.10
267 ('79-'80)	2.45	2.100	3.48
302 ('68-'69)	2.45	2.100	3.00
305 ('76-'85)	2.45	2.100	3.48
307 ('68-'73)	2.45	2.100	3.25
327 ('68-'69)	2.45	2.100	3.25
350 ('68-'85)	2.45	2.100	3.48

LARGE JOURNAL CRANKS

Engine	Main Journal Diameter	Rod Journal Diameter	Stroke
400 ('70-'80)	2.65	2.100	3.75

ONE-PIECE REAR MAIN SEAL CRANKS

Engine	Main Journal	Rod Journal	Stroke
305	2.45	2.100	3.48
350	2.45	2.100	3.48

small-block's tendency to leak at the rear main seal. The one-piece rear main seal design reduced the crank's flywheel mounting flange size, creating a large, round hub sealing surface. This also required a redesigned cylinder block mounting flange, different oil pan and specific damper and flywheel/flexplate.

Since 1986-and-later blocks will soon become more plentiful in recycling yards as cores, it's important to know how this one-piece rear main seal change affects small-block interchangeability. The one-piece seal crank requires the use of a matching 1986-or-later block. However, you can interchange an older 2.45-inch main journal, 2-piece rear main seal crankshaft in these later blocks, and as you can imagine, this complicates things further with a number of different seal adapters, depending upon the style of oil pan you wish to use. The chart "One Piece or Two?" in Chapter 3 deals with the various block, crank and oil pan options. In addition, the one-piece seal cranks also feature a rolled radius fillet in the transition area between the crank rod and main journals and the cheek of the crank.

THE EASY 383

The popular 383 cid small-block package is based on using the 400-style crankshaft in a 350 cid cylinder block. Of course, as you know from our crank chart, the 400 crankshaft is machined with larger 2.65-inch main journals and standard 2.10-inch rod journals and is externally balanced. To fit the crank into the 2.45-inch main journal block, the crank's main journals are turned down to the 2.45-inch spec. Since most of these engines are fitted with used cast cranks, the rod journals are usually turned .010-inch undersize to restore the journal surface. Recently, Callies introduced a new Pro Street forged, mid-priced alloy steel crankshaft that offers the advantages of improved bending fatigue durability and overall torsional strength over either a 1053 forged crank or cast crank, at a reasonable price. Lunati also now offers a 4340 non-twist forging in a number of different strokes at a competitive price. Other companies such as Crower and Lunati also offer a wealth of 4340 forged crankshafts for this and many other stroke combinations.

INTERNAL AND EXTERNAL BALANCE

In addition to all the other small-block crank differences, there's also internal versus external balance considerations.

CRANK TERMS

The world of high performance crankshafts uses a number of specific terms that may not be familiar to the average hot rodder. Below are some of the terms that you may run across in working with or reading about crankshafts.

Indexing—Since crankshafts are a production item built in great numbers, most production crankshafts do not always measure up to their published stroke. Many performance crank grinders offer indexing where the stroke on each of the crank's four throws are measured and machined by offset grinding to the stroke specified.

Offset Grinding—This involves grinding the rod journals undersize and offset to increase or decrease the length of the stroke. For example, a 2.100-inch rod journal can be offset ground down to a 2.00-inch journal size that will increase the stroke easily by .090-inch! This is a small change (just over 9 cubic inches on a 350 for example) but can be used for incremental displacement adjustments.

Twist vs Non-Twist Forgings—When forging a crankshaft, there are basically two ways to make the crankshaft die. In a non-twist crank, all four throws of the crank are forged at the same time. This means the die is more complex and expensive. Proponents claim this design is superior to the twisted forging. Twist forgings use a simplified flat forging die. The crank is then twisted to create the four crankshaft throws.

Welded Cranks—Often, a custom stroke crankshaft can be created by welding additional material to a rod journal and then offset ground to create a longer (or shorter) finished stroke. This procedure can be dangerous since porosity and embrittlement can create stress cracks that could be difficult to detect. Welding can also be used to repair a cracked crankshaft. With the proliferation of high performance steel cranks, there is little reason to consider a welded crank.

Most small-block Chevys are internally balanced engines. This means the torsional damper and flywheel/flexplates are zero balanced. However, the 400 small-block and the '86 and later one-piece rear main seal small-blocks are externally balanced engines. Note the externally balanced torsional balancer (right) compared to an internally balanced ATI friction balancer (left). The ATI is also SFI approved for competition such as NHRA drag racing.

Most small-block cranks are internally balanced where both the damper and the flywheel/flexplate are zero, or neutral balanced. Most enthusiasts are also aware that the 400 cid small-block is an externally balanced engine requiring offset weights on both the torsional damper and the flywheel/flexplate.

What many Chevy fans may not know is that all '86-and-later Chevy small-blocks are also externally balanced engines requiring specific flywheel/ flexplates and torsional damper. This rarely poses a problem with the '86-and-later engines since they also require the specific one-piece rear main seal crank flange flywheel/flexplate. This external balance also differs from the 400 cid style since the torsional damper (often called a harmonic balancer) is externally balanced but the weight is located on the inner hub of the balancer. Externally balanced flywheel/flexplates are easily spotted by the additional weight present on the flexplates and drilled holes in the flywheels. Do not interchange externally balanced flywheel/flexplates and dampers with internally balanced components or the engine will suffer a severe imbalance condition that will shake it right out of the engine compartment!

SPOTTING A GOOD USED CRANK

If it weren't for good used parts, hot rodding might never have happened. Even though almost all used small-block cranks are now made out of nodular iron, they usually survive 100,000 street miles very well. But that doesn't mean you can just yank a used crank out of a leaky old short-block and plunk it into your street small-block. The best way to determine the origins of your crank is to pull it out of the used engine yourself. Often, however, you may find a used crank at a swap meet or in the used parts bin at your local speed shop. Since visually identifying used cranks is dangerous business, measuring the main and rod

BY THE NUMBERS

Chevrolet offers a number of different crankshafts for the small-block Chevy. Of course, Chevrolet is not the only source for high performance crankshafts, but we will limit this chart to just cranks available from Chevrolet. Both iron and steel cranks are listed here. All cranks are the older, two-piece style unless noted as a one-piece seal crank.

PART NUMBER	DESCRIPTION
3951527	3.75-inch stroke, ductile iron crank for 400 engine with 2.65-inch main journal
3941180	1053 steel 350 crank, 3.48-inch stroke, 2.45/2.100-inch journals. Not nitrided, 1182 casting number.
3941184	Same as above crank except nitride heat treated
100501168	4340 steel, raw, unmachined non-twist forging. Accommodates 3.25 to 4.00-inch strokes. The 2.900 main journal diameter can be machined to fit 400 main saddles.
ONE-PIECE SEAL CRANKS	
14096036	1053 steel forging with 2.45-inch main and 2.100-inch rod journals. Used in the 350 H.O. engine PN 10134338

journals and estimating the stroke are the only true tests for a used crank.

Measuring & Inspecting

As evidenced by the chart listing main and rod journal sizes, the best approach is to take some type of precision measuring tool such as a micrometer or dial caliper with you when it's time to purchase a used crank. A 350 crank will often measure just slightly smaller than its 2.45-inch main journal and 2.10-inch rod journal dimensions due to normal wear. Assume that any used crank will necessitate grinding both the main and rod journals. It's also possible that the crank may have already been machined once. Typically, most cranks are machined .010-inch on both the main and rod journals.

Measuring the exact stroke is difficult without a precision tool, but estimating the stroke to ensure the crank is either a 350 or 400 crank can be done by measuring the distance from the rod journal to the main journal. Keep in mind that this distance will be roughly half of the total stroke, so a 350 crank will measure roughly 1-3/4 inches or half of the 3.48 stroke figure. Another way to identify a 350 crank is by its 3932442

casting number. Be aware that the 305 crank can be easily misidentified as a 350 crank since the stroke and journal diameters are identical. See sidebar "The Impostor" to determine how to identify the 305 crank.

Other important areas to check include an overall inspection for blue heat discoloration that could indicate structural damage. Inspect all oil holes for possible stress cracks or damage. The crank snout area should not show evidence of damage, and look closely at the crank keyway slots for indications that the balancer might have widened the slot, which will make repair difficult. Look for nicks or dents in the crank throws that might indicate the crank has experienced structural damage. On the back end of the crank, look closely at the thrust surfaces and oil slinger. High static pressure plate loads or a misaligned torque converter can cause excessive forward thrust that will not only damage the thrust bearings but the crank as well. If the vertical crank thrust surfaces are damaged, look elsewhere for a crank.

Magnaflux—If the crank passes all of these visual inspections and measures out to be the crank you need, have it Magnaflux inspected to eliminate any question of its durability. If you've checked the crank carefully, small stress cracks in or around the oil holes should not eliminate it from consideration. Consult with the shop doing the Magnaflux inspection. Small surface cracks do not automatically eliminate a crank if you intend it strictly for mild high performance street use.

Grooves—Never use a crank whose main or rod journals have been grooved, nor should you use fully grooved main or rod bearings. These grooves merely reduce the surface area of the bearing and overloads the remaining journal area. Cross-drilling a crank is also not necessary and some engine builders believe this process actually will hurt lubrication at very high rpm levels.

In 1986, Chevrolet changed the design of the crankshaft to accommodate a one-piece rear main seal. The standard small-block two-piece rear main seal is illustrated on the left while the one-piece rear main seal crank is on the right. Two-piece seal cranks can be used in a one-piece seal block with the appropriate aluminum seal adapter. However, the one-piece seal crank cannot be adapted to earlier two-piece seal blocks.

This is a Callies lightweight 4340 crankshaft. Note the hollow rod and main journals and lightened counterweights. These lightening efforts trim between 5 and 6 pounds from the crankshaft over a standard 4340 steel crank. This is a two-piece rear main seal crank. The lightweight cranks are usually the most expensive. I don't recommend these cranks for street use.

STEEL CRANKS

In the last few years, there has been an explosion in new aftermarket steel cranks, especially for the small-block Chevy. Callies, Crower, Lunati and numerous others have flooded the market with high quality steel crankshafts. Chevrolet also offers a limited selection of steel cranks, including two raw forgings that can be custom machined to a number of different strokes. Refer to the "It's All in the Alloy" sidebar that indicates the relative strengths of each of the five popular alloy steels used to make crankshafts.

It would probably take a book this size just to detail all the different high performance steel cranks available through all of these aftermarket companies. For example, Crower offers four different styles of 4340 steel alloy cranks in a bunch of different stroke and journal configurations, not to mention 4340 cranks for one-piece seal blocks. These cranks differ mainly in total weight. Lightweight cranks are great for all-out acceleration since accelerating this additional crankshaft weight costs power as rpm rapidly climbs. These lightweight cranks (usually only around 5 to 8 pounds lighter than their heavier cousins) sacrifice ultimate durability for small increases in power and are also very expensive. They are used mainly in drag race, sprint car and NASCAR short track engines and are not necessary nor needed in even high output street engines.

It's crucial that these expensive crankshafts be intelligently combined with the remainder of the bottom end components. For example, spending the bucks for a killer 4340 Callies stroker crank is foolish if you then combine it with stock connecting rods. A bulletproof crank can still be destroyed by a failed stock rod. A wiser approach would be to invest in Callies' less expensive 530A36 Pro Street crank, for example, and then use the remaining money for a set of matching 4340 steel connecting rods. This set would be a much stronger assembly and would offer bulletproof durability for even the strongest street engine. This concept also applies to the entire engine. It makes little sense to invest in this killer rotating assembly if all you're going to do is bolt on a set of stock iron cylinder heads. The money would be better spent on a good set of heads and induction rather than on a NASCAR type bottom end. Crankshafts don't make power—cylinder heads do.

BALANCING ACT

You should always include balancing to the budget for any high performance small-block Chevy buildup. You can get by without balancing an engine equipped with stock rods and replacement forged or cast pistons. These pieces are close to OEM weights, but they are just that—close. A high performance engine means carefully balancing the rotating assembly to eliminate vibration that not only smoothes out engine operation, but is worth small increases in power as well as improved bearing life. I routinely balance

The thrust bearing on a small-block Chevy is located on the last (Number Five) main bearing. The thrust bearing's vertical bearing surface limits horizontal (fore-aft) movement of the crank. Most of this movement is caused by high pressure plate pressures or thrust of the torque converter. Excessive automatic transmission line pressure or a ballooned converter can push the converter forward which can quickly destroy the thrust bearing and/or the crankshaft.

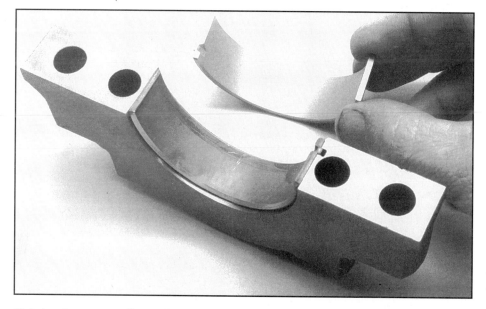

Main bearings are usually supplied with a grooved upper bearing and a non-grooved lower. The non-grooved bearing is always installed in the main cap because this is the bearing half that receives the thrust of the power strokes.

all engines that leave my shop, balanced on our electronic Hines balancer, including both the torsional damper and the flywheel/flexplate that will be used on the engine.

BEARING UP

As with virtually every other small-block component, it pays to invest in the best bearings for your street or race small-block. The main job of an engine bearing is to support the oil film and to allow dirt to embed into the soft outer surface of the bearing without destroying the bearing or the crankshaft. While a soft bearing might be the best for embedability, it would give up too much durability. Therefore, the bearing designer must also make the bearing hard enough to withstand the pressures of high rpm and great torque loads. Most of the new performance bearings on the market now are constructed with three metal layers that start with a steel backing followed by a metal alloy usually consisting of copper and lead overplated by a third, very soft metal Babbitt material. This construction is more expensive but superior to aluminum-backed bearings or production two-metal designs.

Clevite 77's

I prefer to use Clevite 77 bearings in all of my high performance street engines. These bearings utilize this Tri-Metal design that is excellent in creating smooth bearing surface while combining fatigue resistance with conformability and embedability. The Clevite 77 Tri-Metal design is available in a number of different configurations, including .001-inch undersizes that allow you to custom fit specific bearing clearances. Plus, Clevite offers four different construction styles to meet various high performance needs. Among these, the Micro-Babbitt and DeltaWall bearings are not intended for high performance street use. The two more readily used Clevite bearings are the H and P-series bearings that offer different eccentricity contours to allow the engine builder to customize his bearings to create an optimal wear pattern. More information on these bearings can be found in the Clevite 77 bearing catalog.

Inspection

When tearing an engine down for a basic rebuild or freshening, bearings are an excellent indicator of potential problems. This is why it is critical to closely inspect every bearing in an engine when it is disassembled for service. It is beyond the scope of this book to detail all the possible different bearing failures and their corrective action. However, Clevite 77 does offer a free Analysis and Correction of Bearing Failures booklet just by writing to Clevite c/o JPI Transportation Products, Inc., 325 E. Eisenhower Parkway, Suite 202, Ann Arbor, MI 48108-3388. Please include your name and complete mailing address.

The most important point in Clevite's booklet is that dirt, improper assembly and misalignment represent 71 percent of all premature bearing failures. Often a bearing fails because it has been subjected to a combination of problems that led to its failure. When you discover a failed bearing, the key is to evaluate why the bearing failed and take the necessary corrective action to prevent a repeat failure. Of course, Clevite 77 bearings are just one of many quality bearings available. Just be aware of what you're buying and check before you blindly accept an unmarked box of engine bearings.

TORSIONAL DAMPERS

For most small-block street applications, the stock damper will work just fine. Be aware that over the years, Chevrolet has changed the position of the timing marks on the damper. Dampers produced from 1969 and later locate the TDC mark 8 degrees further advanced than dampers produced before 1969 where the TDC mark is aligned with the keyway slot. Mixing and matching dampers and timing indicators will result in erroneous timing figures. If you are unsure whether your damper is an early or late style, the best procedure is to check your TDC mark with a piston stop in the Number One cylinder.

OEM-style dampers are acceptable for even high performance street use as long as the elastic rubber separating the inner and outer hub appears to be in good shape. If the rubber looks cracked, is split or has pushed out from between the hub, find a better damper. Notching timing

Installing a damper properly requires a balancer installation tool. Do not beat the balancer on with a hammer and a block of wood. This Crane balancer installation tool uses a threaded insert into the crank snout and a bearing fitted to the outside tool to reduce friction.

marks in the damper can help with power timing and some shops offer this as an extra-cost service.

If your killer small-block can benefit from an aftermarket high performance damper, I prefer to use the ATI friction damper. The damper utilizes springs and a friction material that dampens the torsional frequencies generated by the crankshaft at various rpm points.

CONCLUSION

Stock ductile iron crankshafts do an admirable job for the production small-block. They will easily last 100,000 miles, undergo a regrind and offer another 100,000 miles when treated

properly. Even for street performance work, there is little need to step up to stronger steel crankshafts unless high rpm or sustained high output is planned. In most street applications, a properly machined production ductile iron crankshaft with blueprinted bearing clearances will withstand even the heaviest right foot. But if you plan on really leaning on an engine, investing in a good high performance steel crank and matching connecting rods is an excellent foundation for a killer small-block. Typically, this decision to buy a 4340 steel crank is driven more by the limitations of your wallet than by your desire to invest in the best parts. ∎

CONNECTING RODS 5

Connecting rods may be the most obscure part in the small-block Chevy. Connecting rods may not be responsible for increasing power by a magnitude of 10, but ignore the rods when it's rebuild time and the result could be costly. For most small-block Chevy street engines, properly rebuilt stock rods with high quality rod bolts will survive for hundreds of thousands of flawless miles. Despite their pedestrian position in the power chain, it's best not to ignore the realities of putting the performance twist to stock rods. We'll also take a look at some of the steel rod aftermarket options available to the performance builder.

Small-block rods come from the production foundry in two basic lengths, the 5.7-inch rod and the 400-style 5.565-inch length. Length is determined by measuring the center-to-center distance between the big end and wrist pin end. While stock rods have an excellent reliability record, there are still a few rebuild basics that will help you avoid problems. Among the 5.7-inch rods, they break down into '67 and older rods used on the small 2.00-inch rod journal crankshafts such as the 283, '67 302 and '67 and earlier 327 engines. All '68 and later small-blocks use the more common 2.100-inch rod bearing journal diameter. These small journal rods also use a 11/32-inch rod bolt while the 2.100-inch rods use a larger and stronger 3/8-inch rod bolt. Since the small journal rods are not

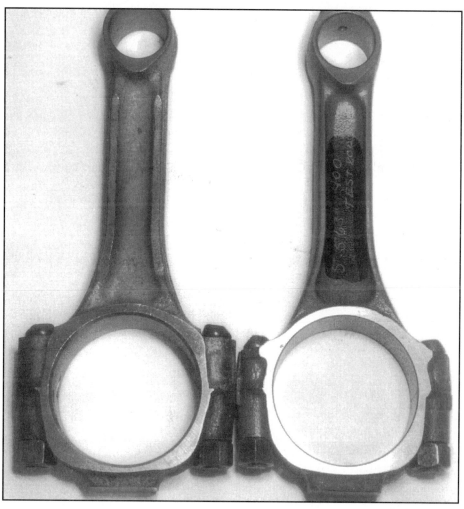

The two most popular small-block Chevy production rods are the 350 5.7-inch long rod (left) and the 400 5.565-inch rod (right). Both rods require a 2.100-inch rod journal.

nearly as common as the larger, later rods, we'll leave the details of those parts to the musclecar restoration fans.

HIGH PERFORMANCE STOCK RODS

It's no great revelation that stock rods perform admirably in mild street small-

blocks. But there's more to stock rods than just bolting them into your engine, especially if they are a part of a high performance small-block. While it may be tempting to buy a set of rebuilt stock rods from one of the volume discount houses, these inexpensive rebuilt rods typically extract a higher, hidden cost in

The production "pink" small-block rod (bottom) is merely an upgraded production rod that has been Magnafluxed and shot peened. I prefer to use the Oliver 4340 steel rod (top) for serious performance applications.

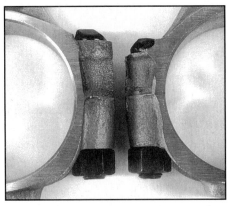

The 400 rod bolt (right) is noticeably shorter compared to the 350 rod (left). This extra clearance is used to clear the cam and block due to the 400's longer stroke. When using a 350 rod with a 400 crank in a 350 block for a 383, you may have to trim the rod bolt and block to produce sufficient clearance.

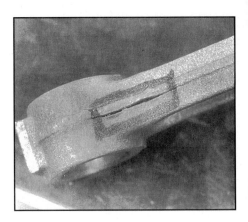

You wouldn't need Magnafluxing to see this crack! This is an obvious example of where production rods often fail. Stock rods also often crack around the rod bolt area.

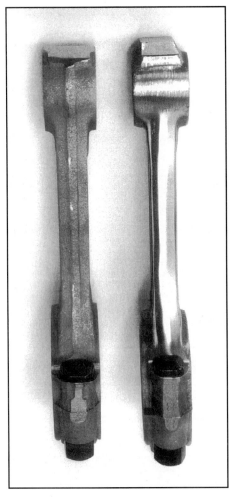

In an effort to reduce the chances of stress fractures on the surface of the rod, many hot rodders will polish the rod beam. This is a perfectly acceptable procedure although not a necessity. If you decide to polish the beams, always grind in the vertical direction, never sideways. It's recommended that you shot peen the rods after polishing.

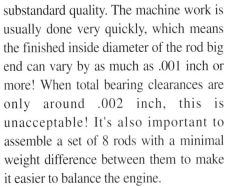

substandard quality. The machine work is usually done very quickly, which means the finished inside diameter of the rod big end can vary by as much as .001 inch or more! When total bearing clearances are only around .002 inch, this is unacceptable! It's also important to assemble a set of 8 rods with a minimal weight difference between them to make it easier to balance the engine.

Bearings are often blamed for improper clearances when the problem exists with a too-small or too-large housing bore. I spec all rods at 2.1250-inch for all 2.100-inch journal rods.

Selection & Basic Prep

Many street engine builders prefer to use Chevy's "pink" small-block rods, which is merely a Magnafluxed and shot-peened version of a 1038 steel production connecting rod. This is certainly a better piece than a stock rod, but consider the alternatives. With Chevy's pricing structure, it may cost only a few dollars more to step up to a much higher quality aftermarket 4340 forged steel connecting rod. Of course, a good compromise would be a set of used pink rods that are obtained for a reasonable price. Another alternative is the Chevrolet Bow Tie rod, which is also a 4340 forged steel piece. Unfortunately, the Bow Tie is much heavier than its aftermarket alternatives such as Oliver, Crower, Callies and others plus it costs more money. Since weight is a factor, I prefer to use aftermarket rods rather than the Bow Tie forgings.

The first step after choosing a stock rod is to have them Magnafluxed. This process is used to find surface cracks in items like crankshafts, heads and rods. Before you invest money into used rods, make sure they are free of cracks. Small-block rods typically fracture either in or near the vertical parting line (see photo) or around the rod bolt area of the big end. If the rods pass the Magnaflux test, another wise investment would be to treat the rods to shot-peening. This is a surface

ARE YOU CLEAR?

There are a number of important clearances relating to connecting rods, especially if you are building a stroker small-block. There is little room between the rod bolt and the camshaft, especially with rods 1, 2, 5 and 6.

LOCATION	CLEARANCE (inches)
Rod bearing	.002 - .0025
Rod side clearance	.009 - .013
Rod bolt to cam	.050
Rod bolt to block	.050

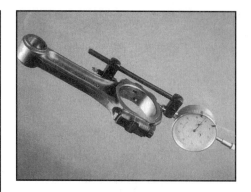

Torquing stretches any bolt to establish a clamping load. The best way to ensure proper bolt stretch is to use a rod bolt stretch gauge such as this Childs & Albert gauge. Too little bolt stretch could allow the nut to back off. Too much stretch will damage the bolt. Both of these situations will lead to rod bolt failure.

preparation technique that costs usually less than $50 that reduces the chance of stress risers forming on the surface of the rod. Some enthusiasts also polish their rods by grinding the surface of the rod beams smooth. While this isn't necessary, it does reduce the risk of stress risers that can lead to fractures.

Rod Bolts—The most important step you can take to improve the strength of a production rod is to invest in a set of high quality rod bolts. It is the bolt's job to keep the two-piece big end of the connecting rod together. American Racing Products (ARP) has established itself as one of the premiere bolt manufacturers for high performance engines and they produce a number of excellent rod bolts for the small-block. I prefer to use ARP's upgraded 3/8-inch rod bolt as opposed to the basic ARP 190,000 psi rod bolt. This higher quality bolt is not that much more money and offers additional insurance against failure.

The details of rebuilding the big end of the rod will be covered in Chapter 7.

However, it's important to mention that new bolts should be carefully installed to prevent cocking the bolts by "catching an edge" and perhaps wedging metal under the rod bolt head as it is installed. This means you don't beat them in with a hammer on the edge of a vise. To prevent damage to the bolt or the rod, a gentle radius between the bolt head surface and the edge of the rod is critical. This radius

ARP offers both a 190,000 psi high quality rod bolt along with the latest Wave Loc bolt that offers both increased 220,000 psi strength and a waved body that replaces the knurling in the rod bolt body. The top bolt is a stock 350 rod, the middle bolt is an ARP and the bottom is a shorter ARP bolt for the 400 rod.

Torquing rod bolts in the engine is a critical step. Torque stretches the bolt to a length predetermined by the manufacturer. The bolt torque spec accounts for bolt stretch, the friction between the threads and also between the nut and the rod cap. This is why you should follow the rod bolt manufacturer's lubrication recommendations carefully. Using moly lube in place of oil and torquing the rod bolt to the same spec creates less friction which applies more stretch to the bolt. Excessive stretch damages the bolt and could cause a failure! Inadequate torque will also allow the nut to back off with disastrous results.

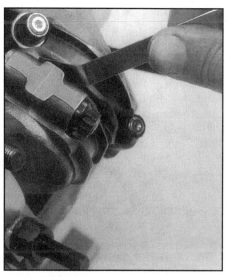

Side clearance is important for many reasons. Obviously, there must be sufficient clearance between the rods to prevent galling and possible failure. Side clearance also affects the amount of oil that exits the rods. Excessive rod side clearance can contribute to excessive oil sprayed onto the cylinder walls, making oil control more difficult for the piston rings.

BEAM WIDTH

Most hot rodders don't know that Chevrolet narrowed the beam width of the small-block Chevy connecting rod beginning approximately with 1989 and later connecting rods. The following chart illustrates the difference between earlier small-block production rods, the narrow beam rods and an aftermarket Crower rod. Note that the Crower rod is not only heavier but substantially wider in the beam. This is a case where strength warrants the additional weight. I do not use the later, narrow rods in any performance application. I have seen occasional rod failures in some of the new LT-1 engines that use this rod. Also notice that although the beam is narrower, the rod is just as heavy, with the weight placed in the big end which actually places more stress on the rod at higher rpm. All of my late-model engines, such as the LT-1 350 and 383 packages, replace these narrow beam rods with either earlier, stronger production rods or optional Oliver 4340 forged steel rods.

CONNECTING ROD	ROD WEIGHT (grams)	BEAM WIDTH (inches)
350 Rod ('89 & earlier)	600	.570
350 Rod ('89 & later)	602	.505
Crower Sportsman (5.850-inch)	712	.685

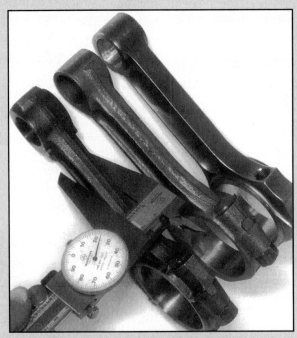

Beginning around 1989, Chevy narrowed the beam width of the production rod by roughly .050-inch. This thinner, weaker rod has contributed to a few high rpm rod failures in strong small-blocks. That's why I do not use these narrow beam rods in any performance small-block. From left to right is a .620-inch thick Oliver forged steel 4340 rod, the standard beam, .575-inch thick, 5.7-inch Chevy rod, and the weaker, narrow beam .510-.520-inch thick rod.

also prevents stress risers.

AFTERMARKET RODS

Hand in hand with the proliferation of aftermarket high performance crankshafts for the small-block Chevy is the tremendous number of high-performance connecting rods. Companies like Oliver, Callies, Crower, Manley, Childs & Albert and even Competition Cams and many others offer a tremendous variety of connecting rods from which to choose. Deciding upon a set of high performance rods will probably be determined by your choice of crankshaft as anything else. If you choose to remain with a stock cast iron crank for a basic street engine, there's little necessity for a killer set of rods. However, the choice of a steel crank should at least warrant consideration of upgrading the connecting rods as well. An good compromise would be a steel crank in the 5130 steel range and matching the crank with a set of quality 4340 steel rods. This discussion will be limited to 4340 steel connecting rods since I do not recommend aluminum rods for the street and titanium rods are not only unnecessary but prohibitively expensive. Virtually all of the aftermarket steel rods are forged from 4340 steel and offer high strength bolts in both 3/8 and 7/16-inch diameters.

Rod Length

Besides design differences in the many aftermarket rods, the basic choice comes down to rod length. As discussed in the Rod Length sidebar, a longer rod is typically a good idea for improved durability that can also pay off in power dividends as well. Especially with stroker crankshafts, a longer rod is a worthwhile addition. I use Oliver rods in engines that opt for an aftermarket rod although you may want to do a little comparison shopping of your own. Manley now offers a new, inexpensive 4340 rod that looks attractive.

Just to make the selection process a bit

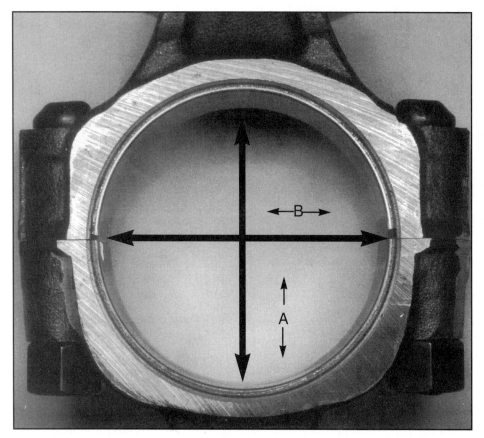

Bearings are designed with a certain amount of eccentricity. This means the i.d. of the bearing is greater at the parting line than it is in the vertical plane. This prevents the bearing insert parting lines from digging into the crank journal and accounts for big end elongation under power. As shown, the vertical distance A is less than the horizontal distance B. Always measure bearing clearance in the A distance.

Measuring the inside diameter of the bearing in the rod means torquing the bearing in place. The only acceptable way to do this is to clamp the rod in a rod vise with both the rod cap and big end body properly supported. This prevents twisting the rod when applying torque to the bolts.

BEARING UP

Bearing clearances are determined by subtracting the crankshaft journal diameter from the inside diameter (i.d.) of the rod bearing. That part's simple. Bearing clearances can vary even when the crank journals are all exactly the same because the inside diameter of the rod (minus the bearing) varies. Listed below are Chevrolet production ranges for both small and large journal rods. This i.d. plays a significant role in establishing bearing clearance. When the rods are held to a common i.d. with quality machine work, bearing clearances will be more consistent. I've also included my recommended finished big end dimension that is used to maintain proper bearing clearances.

ROD BIG END DIMENSION (inches)		LINGENFELTER SPEC (inches)
2.00	2.1247-2.1252	2.1250
2.10	2.2247-2.2252	2.2250

more complicated, most rod manufacturers offer more than one style of rod from which to choose. Crower offers five or six different small-block connecting rods in either 4340 forged or billet constructions. These rods vary in weight, strength, and "profiling" which trims the rod to clear the cam and block in stroker configurations. For street engines, the least expensive 4340 steel forged rods offered by any of these companies offer dramatic increases in connecting rod durability.

Weight, rod bolt clearance, and price are all important considerations within a specified rod length. Keep in mind that there is an effective limit to rod length determined by piston compression height. Long rods require custom pistons to accommodate the added rod length. Many companies like Callies, Crower, Childs & Albert and Lunati offer matched kits that include rods, pistons and even crankshafts to make the selection process a little easier.

ROD LENGTH TO STROKE RATIO

The length of a connecting rod does have an affect both on power and

Rod bearings operate in a difficult environment subjected to heat, pressure, dirt and crud. This bearing was pulled from a new engine. Obviously, dirt and perhaps excessive heat have contributed to bearing distress.

Stock cranks have a rather narrow fillet radius between the crank journal and the cheek of the crank. High performance cranks employ a wider fillet for greater strength. In these cranks, a wider chamfer rod bearing is needed to prevent scuffing the bearing against the crankshaft. This is an important test with any aftermarket crankshaft. All the major bearing manufacturers offer wider chamfer bearings for these applications.

durability. The small-block Chevy is especially limited with its rather short 9.025-inch deck height, making rod length an important consideration even for street engines. As you've seen in the chapter on displacements, the 383, 406 and 420-inch small-blocks typically employ strokes of 3.75- to 3.875-inches. Increasing stroke decreases the available room for the connecting rod even when the wrist pin is moved closer to the ring package. This situation is most evident with the 5.565-inch long rod used in the production 400. This .135-inch shorter rod combined with the 400's 3.75-inch stroke tends to increase rod angularity which increases piston loading on the thrust side of the cylinder wall. In essence, this angle tends to shove the piston into the cylinder wall on the thrust side. Shorter rods also tend to accelerate the piston away from TDC more quickly.

One way to look at rod length is to compare it to the engine's stroke. This is because rod angle is affected by both rod length and stroke. In other words, a 5.7-inch rod creates a much more acute angle in a 3.75-inch stroke engine than it does with a 3.00-inch stroke. This relationship is best expressed by dividing the rod length by the stroke, creating a rod-length-to-stroke (R/L) ratio. For example,

a 350's 3.48-inch stroke with a stock 5.7-inch rod length generates a 1.64:1 R/L ratio. A stock 400 drops the ratio to 1.59 with its longer 3.75-inch stroke and 5.565-inch connecting rod. Conversely, a 302 with its 3.00-inch stroke and 5.7-inch rod creates a more favorable 1.9 L/R ratio.

Like many engine builders, I personally favor a long rod relative to the stroke to reduce frictional losses and bore wear. This is why spending the extra money to put 5.7-inch long rods in a 383 using a 400 crank (requiring more expensive pistons) is a good bet. The least expensive approach is to use the stock 5.565-inch rods, but this tends to place additional pressure on the cylinder walls which also creates additional friction. Longer 5.7-inch rods are a better choice. The worst case small-block is the 420 with its 3.875-inch stroke. Even with 5.850-inch rods, the L/R ratio is a short 1.51:1. While this is less than ideal, my 420's are usually aimed at maximizing torque and not high rpm horsepower. Therefore, the shorter R/L ratio isn't a critical problem. In order to create a more comfortable 1.64:1 R/L ratio, the rod would have to be 6.355 inches long! A more typical ratio would be 1.58:1 with this engine using a 6.125-inch rod. Even

this would require a tall deck block like the Oldsmobile Rocket block.

BEARINGS

Rod bearings are probably even more ignored than connecting rods. Most hot rodders just check clearances, plug the bearings in and forget about them. But there's more to bearings than just clearances. In fact, rod and main bearings are often blamed for improper clearances when the housing bore (either rod or main) or the journal diameter is the real culprit for too-loose or too-tight clearances. If you spend some time measuring bearing shell thicknesses, you'll discover that quality bearings often hold their tolerances to well within .0001-inch, which is often difficult to measure accurately!

Clevite 77

There are a number of very good manufacturers producing bearing inserts for the small-block Chevy. Like most engine builders, I have established a solid foundation of experience with Clevite 77

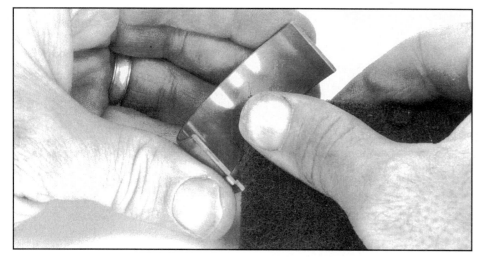

According to bearing engineers, there is no reason to ever Scotchbrite engine bearings. This is a stone age trick used by engine builders who thought they knew more about bearings than the engineers who designed them. If you insist on no flash coating, Clevite now offers non-flash-coated bearings in the H-series line.

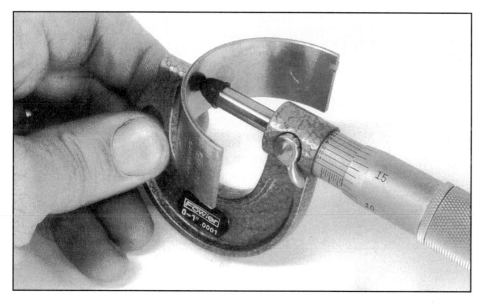

The best way to measure bearing clearance is to measure the journal diameter, the rod big end i.d. and the bearing shell thickness. You can then compensate for variations between these three variables to establish a consistent bearing clearance. Bearing shells can be measured with a ball mic. Or you can purchase this slick ball adapter for a standard 0 to 1-inch mic from the Grainger catalog for around $6.00. Don't be surprised to discover quality bearing shells maintain a thickness consistency within .0001-inch.

bearings and use them almost exclusively. Ask other engine builders and their preferences may be different. That's not right or wrong, just different.

Clevite 77 bearings are constructed in a Tri-Metal configuration. As outlined in the crankshaft chapter, they consist of a steel backing with a cast, copper-lead alloy layer overlaid with a precision electro-plated babbitt top layer. This is a typical configuration for high

performance bearings offering the best compromise between embedability, which handles small bits of dirt, and fatigue resistance to handle the high loads, especially in a high output engine.

Clevite actually offers four different small-block bearing types, although I use only two of the four for street engines. For mild street engines, the P-series bearings offer the best surface for use on iron crankshafts since they are more

compatible with the stock crank. For steel crank engines, I prefer the H-series bearing, which is slightly harder. The H-series bearings also come with a larger chamfer that matches the typically larger radii used on high performance crankshafts. This extra chamfer clears the wider crank radius to prevent the bearing from hitting the crank fillet.

When disassembling an engine, you should also pay attention to the wear pattern of the rod bearing. A proper wear pattern would indicate wear from about 2/3 to 3/4 of the bearing surface. A too-wide wear pattern could indicate too little eccentricity in the bearing. The P-series bearing adds a slight amount of eccentricity to the bearing thickness to compensate. Conversely, a narrow wear pattern (wear concentrated in the middle of the bearing) indicates too much eccentricity so an H-series bearing might be a better choice. While these are somewhat esoteric concerns, it's best to know the differences in bearing construction in order to make the best choice.

If you don't retain anything else about bearings from this chapter, at least remember that using Scotchbrite to prepare rod or main bearings is definitely not recommended! This procedure was used in the past when engine builders believed a bearing's flash coating would tend to ball up on the bearing and create problems. Clevite 77 engineer Bob Anderson recommends that the flash coating be left intact on the bearing. If you insist on a bearing without the flash coating, H-series bearings currently come without the coating, giving the bearing a dark, non-reflective appearance.

Regardless of whether you choose a stock connecting rod or a killer aftermarket piece, make sure they have received the same attention and quality machine work as the rest of your engine. Connecting rods are one area where you can least afford to be penny wise and rod foolish. ∎

6 PISTONS AND RINGS

Tuning the induction, exhaust and cam timing is all an effort to maximize turning the air and fuel mixture into heat and power. Pistons are what make it all happen, and as with everything else in a small-block, there are plenty to choose from. There are literally hundreds of types of small-block pistons available from many sources for all of the different bore/stroke/compression ratio/ring combinations that have evolved with the Mouse motor of the years.

TYPES OF PISTONS

Despite the pile of aluminum slugs that we would create if we just chose one example of each of the different small-block pistons, there is sense to be made out of this madness. There are generally three distinctly different types—cast, hypereutectic and forged—that make up the world of performance pistons. From there, things get a little crazier with variations in compression heights, ring thicknesses, ring placement and a few other details that we'll get into as we move along. By the time you're done with this chapter, you'll discover there's a lot more to pistons that just squeezing that air and fuel so it'll burn.

Cast Pistons

In the no-frills, work-a-day small-block world, the cast aluminum piston plays the part of Mr. Dependable. Examples of the breed are Badger, Silv-O-Lite and

When it comes to pistons, the small-block Chevy engine builder has many choices. There are literally hundreds of types of small-block pistons for all of the bore/stroke/compression ratio combos that have evolved over the years.

Zollner. But add compression, nitrous and/or rpm and cast pistons are destined for a short life span. As the name suggests, cast pistons are poured into a mold, cooled, machined and sent on their way. This quick manufacturing process lends little strength and as a result, detonation and high rpm are two moves guaranteed to ruin a cast piston's day.

Cast pistons are often fitted with a steel thermal expander that is installed when the piston is still hot. This "stretches" the piston when it's cold to maintain the same piston-to-wall clearance as it heats up. This is why you can install cast pistons with tight .0015-.002-inch clearance without scuffing. Cast pistons are also often built with what is called piston pin

Pistons come in three aluminum alloy configurations—cast (left), hypereutectic (middle) and forged (right). Strength is directly proportional to cost. In other words, cast pistons are the least expensive but offer only marginal strength. Forged pistons are by far the best and offer the best durability of the three.

offset. A typical offset would be .060 inch where the pin is closer to one side of the piston that the other. This is used to ensure quiet operation, especially during cold startup when the clearances are the greatest. Some hot rodders using cast pistons will reverse the piston pin offset which can be worth a small amount of power. As you can imagine, the pistons will slap around a little when the engine is cold.

Hypereutectic Pistons

A few years ago, a number of aftermarket companies began experimenting with high silicon content cast aluminum pistons called hypereutectic (pronounced hyper-u-tec-tic). The concept behind these pistons was to create a stronger cast piston that could withstand the higher rpm and cylinder pressures of a street performance engine. Two companies that produce hypereutectic pistons are Sealed Power and Silv-O-Lite under the Keith Black name.

The term hypereutectic is derived from the amount of silicon added to the aluminum alloy. These alloys come in three different categories—eutectic, hypoeutectic and hypereutectic. The saturation point of silicon in aluminum is the eutectic point at 12 percent. Alloys below this point are hypoeutectic and alloys above it are hypereutectic. The Keith Black hypereutectic pistons are cast with a 16 to 18 percent silicon content which dramatically increases strength and wear resistance. The silicon also acts as a temperature barrier, which means it tends to retain more heat in the combustion chamber than other cast pistons yet resulting in roughly 15 percent less thermal expansion over conventional alloy pistons.

This heat also has an affect on piston ring end gap clearance. Silv-O-Lite recommends significant increases in compression (top) ring end gap to accommodate the heat transferred by the top ring into the cylinder wall. For example, on a 4.00-inch bore 350 with a

KB piston, Silv-O-Lite recommends a .026-inch top ring end gap compared to typical ring end gaps for a normally aspirated street engine of .018 inch. This increased end gap sounds excessive, but according to Silv-O-Lite engineers, this cold end gap is reduced down to a hot running end gap similar to a non-hypereutectic cast or forged piston. Second ring end gaps are not affected by the heat so they do not require larger end gaps.

Forged Pistons

If you plan to build a hot street engine of any kind, forged pistons are the only intelligent choice. Since the small-block is the performance parts volume king, this volume allows the engine builder to purchase parts at excellent prices leaving little excuse for saving money by purchasing cast pistons instead of forgings.

Forged pistons are created by packing aluminum alloy into a rough slug by the enormous force of a huge press. This

There are two basic types of dished pistons. Dyno testing has shown that if you need a dished piston to create a certain compression ratio, the half-dished or reverse dome piston is superior since the flat portion of the piston combines with the cylinder head to create a quench area that improves combustion activity.

A quick way to increase compression is with a domed piston (left). Flat-top pistons (middle rear) are perhaps the most common since they work well with relatively large chambers for streetable compression ratios. Dished pistons (middle front and right) increase the effective combustion volume, decreasing the static compression ratio.

pressure creates a much denser grain structure that generates a stronger piston without additional weight. Besides all the other variables of compression ratio, compression height, ring thickness, ring land placement etc., weight is a major factor in piston selection. Obviously, a lighter piston is an advantage, especially in high rpm operation.

CHOICES, CHOICES...

The following chart lists the various advantages and disadvantages of the three different piston types. This is a quick overview but will give you an idea of each piston's relative strengths and weaknesses.

CAST

Advantages: Inexpensive, quiet, lightweight, tight clearances
Disadvantages: Breaks easily under detonation; few options for compression and rod length

HYPEREUTECTIC

Advantages: Stronger than cast; improved choices for compression and rod length; less expensive than forged; reduced heat transfer.
Disadvantages: Not easily modified

FORGED

Advantages: Tremendous options for compression, rod length, ring design, and clearances; Strongest and most durable; Best weight/strength ratio; Widest range of prices.
Disadvantages: Usually most expensive

FLAT, DISHED OR DOMED?

Perhaps the first question addressed once bore and stroke are considered is whether to run a flat, dished or domed piston. The trend today is moving away from large combustion chambers, which means that flat-top and dished pistons are now a reality even at street compression ratios of 9 to 10:1. For example, the new LT1 small-block features a 10.5:1 compression ratio with a small 54cc chamber and a flat-top piston! Given all the different possibilities, it's important to look at all three styles.

Flat-Top

The most common small-block piston head configuration is the flat-top. Stock replacement TRW and Sealed Power pistons come with four symmetrical valve reliefs cut into the head for piston-to-valve clearance. Since these pistons have no pin offset, the piston can be used interchangeably in any cylinder, making it easy for the manufacturer since it requires only one mold. The disadvantage

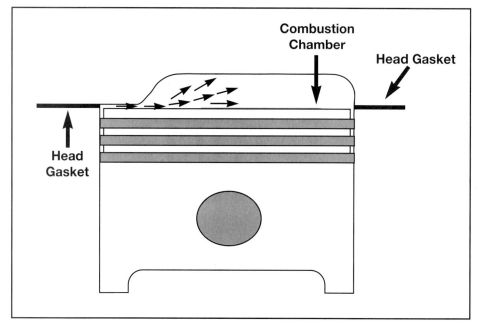

This is a cross-sectional view of how the quench area of the chamber pushes the air-fuel mixture out into the combustion chamber. This action creates turbulence that more thoroughly "homogenizes" the mixture and improves combustion that makes more power. A quench area, which also is the piston-to-head clearance, of .040-inch is achievable with a .002-inch deck and a .038-inch gasket. Some engine builders tighten this figure using a good steel rod and tight piston-to-wall clearance. These steps prevent piston rock at TDC to keep the piston away from the head.

Many enthusiasts don't realize that lightweight components like pistons can improve performance. Heavy replacement style pistons (left) require more power to accelerate as opposed to the lightweight JE piston (right). Trick pistons aren't necessary for a street engine, but will be necessary when building a stroker engine such as a 377 or 420.

MEASURING PISTONS

Each piston manufacturer specifies the exact place its pistons should be measured to establish the proper piston-to-wall clearance. Because pistons are eccentric, measuring the piston diameter in the proper place is critical to establishing the correct piston-to-wall clearance. Below, we've listed some of the more common piston manufacturers and where they specify piston diameter should be measured. As you can see, this position varies dramatically. We have compiled this data to illustrate the differences between piston manufacturers. Since piston design is an ever-evolving science, we suggest that you check with the manufacturer of the pistons you're using before measuring your pistons. With the exception of the Keith Black and Speed-Pro hypereutectics, all are forged pistons. All measurements must be taken on the piston skirt.

PISTON MANUFACTURER	POSITION
JE	.500 inch above bottom of skirt
Keith Black Hypereutectic	Level with bottom of balance pad
Ross (A design)	Wrist pin centerline
Ross (B design)	Just below bottom of wrist pin
Speed-Pro	Wrist pin centerline
Speed-Pro HT-16 Hypereutectic	Level with bottom of balance pad
TRW	Wrist pin centerline
Wiseco	1.300 inch below oil ring groove

to this universal design is that it gives up a few cc's of volume to the two valve reliefs that are just along for the ride. Custom pistons like Ross, JE and others offer two valve relief "left" and "right" pistons with specific intake and exhaust-sized valve reliefs that don't sacrifice this volume. Of course, you must make sure you install these pistons correctly!

Dished

Dished pistons are a little more complicated. At first, all dished pistons were created with a full dish that spanned almost the entire top of the piston. The depth of the dish determined the ultimate compression ratio with more volume reducing the static compression. But experimentation proved that creating a "half-dish" or D-shaped dish had a positive effect on power. This is due to the quench or squish design of the small-block Chevy cylinder head.

QUENCH AREA

If you look at the combustion chamber of any small-block head, the chamber only accounts for a little more than half of the diameter of the bore. The remaining portion of the chamber is flat. With a flat-top piston, the distance between the top of the piston and the flat portion of the head is defined as the quench area. This distance varies

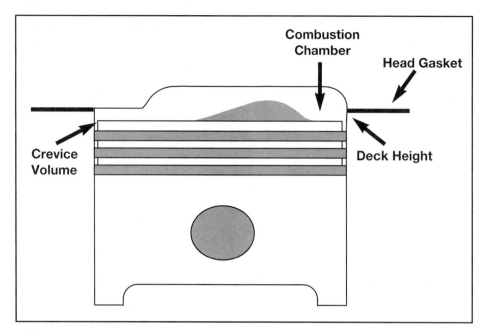

This cross-section drawing illustrates the relationship of the piston top, deck height, head gasket thickness and chamber size to compression ratio. The best way to account for all the crevice volumes (the area above the top ring, and/or valve reliefs on flat top pistons) is to mock up one cylinder with the piston exactly .100-inch down in the bore. Compute this volume as if it were a true cylinder .100 inch tall, then compare the computed volume with the measured volume. The difference is the crevice volume. Add this volume to the combustion chamber volume for an accurate compression ratio.

depending upon the deck height of the piston (positive above the deck or negative below the deck) and the head gasket thickness. A performance piston-to-head clearance is usually between .040 and .045 inch. This quench area is too small to support combustion, serving instead a more useful function. As the piston approaches Top Dead Center (TDC), this tight area tends to squirt or squish the air/fuel mixture in this area into the open combustion space. This violent action produces turbulence that acts like a very efficient blender. This blending action helps homogenize the air-fuel mixture and supports more efficient combustion.

Dyno testing has proven that improving the piston-to-head quench area from .060 inch to .045 inch will improve power even when the compression ratio remains the same. It is even possible to optimize the quench and increase compression without suffering detonation problems! This is because improving the homogenization of the mixture in the chamber reduces the tendency for lean areas in the chamber to promote detonation. The ideal combination is to run a .005-inch negative deck height along with a composition head gasket with a compressed thickness of .038 inch to produce a .043-inch piston to head clearance for a steel rod engine. Optimizing the quench will often reduce the required ignition lead, further reducing the possibility of detonation while also reducing unburned hydrocarbon exhaust emissions. Just for the record, if you are going through this same drill with a domed piston, make sure you still retain a safe piston-dome-to-head clearance of at least .040 inch.

Domed pistons are the classic small-block piston design. Chevrolet built thousands upon thousands of production small-blocks with domed pistons back in the lost days of 103 octane pump gas. Domed pistons are usually reserved for race engines these days for a number of good reasons. Since it's possible to generate well over 11:1 compression with flat top pistons, it makes little sense to stuff a dome into a large chamber to accomplish the same goal. Conventional wisdom also claims that domes merely get in the way of an efficient combustion process and can lead to reduced power even with higher compression. This is especially evident with pistons that require a "fire slot" in the dome to give the spark plug sufficient area to create a flame front. Domed pistons also increase weight at the very top of the piston, which increases the g-forces on the piston. The best way to reduce this weight is to opt for "hollow dome" pistons where the dome area has been relieved to reduce the weight.

These parameters establish the basic configuration of not only forged pistons, but cast and hypereutectics as well. We'll now move on to a closer look at the variables dictated by the specific needs of the engine builder.

COMPRESSION RATIO

Compression is a favorite power play. Everyone knows that squeezing the inlet air and fuel harder will result in more power. But as is the case with most "more is better" theories, street engines have limitations. The first and foremost limitation is the quality of today's pump gasoline. The best premium fuels now are limited to 92 and sometimes 94 octane. Based on this, most enthusiasts will tell you that 9.0:1 is the effective limit for this fuel. As you might guess, this isn't always true. Variables that affect this include iron or aluminum cylinder heads, camshaft duration, engine temperature, rpm, inlet air temperatures and others.

My experience has shown that street engines will live comfortably at 9.2:1 with relatively short duration camshafts with iron cylinder heads. Change to aluminum heads that reduce effective cylinder pressure with greater heat transfer and the static compression ratio can be bumped a full point to around 10.2:1. This illustrates the need to make solid decisions about every component you will use before you buy.

It's also a wise move to not rely on rumors concerning the compression of a certain piston. The only way to know the true compression ratio of your engine is to compute it after measuring all the volumes. Variables include the deck height, various crevice volumes of the pistons (such as the area between the top ring and the piston top), gasket compressed thickness, combustion chamber volume and, of course, the bore and stroke of your engine. We will not detail how to calculate compression ratio here. That can be found in Chapter 17, *Blueprinting*.

COMPRESSION RATIO

COMBUSTION CHAMBER VOLUME (cc's)

		64	68	72	76
PISTON	-22	8.49	8.20	7.92	7.67
(cc's)	-13	9.26	8.90	8.57	8.27
	-6	9.97	9.55	9.17	8.81
	0	10.69	10.20	9.76	9.35
	3	11.09	10.56	10.08	9.65
	11	12.35	11.68	11.09	10.56
	13	12.72	12.01	11.38	10.82

NOTE: This chart is based on a 4.030 bore and a 3.48 stroke with a .015-inch deck height and a .038-inch compressed head gasket thickness. The deck height figure is actually about .010-inch less than a stock small-block's typical .025-inch deck height. This shorter deck height increases compression just like a thinner head gasket or a smaller combustion chamber. This tighter quench also improves combustion efficiency. The total quench is still .053-inch, .013-inch more than the .040-inch figure. Changing the deck height by .010-inch will change the static compression ratio by .28 of a ratio. For example, our chart says a 76 cc chamber head with a pure flat top piston (zero piston cc) combination is 9.35:1. Cutting the deck height another .010-inch would pump the compression by .28 to 9.63:1.

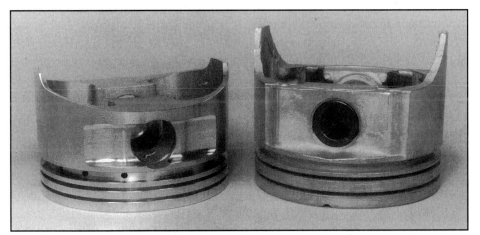

Compression height is defined as the distance from the centerline of the wrist pin to the top of the flat portion of the piston. Increasing stroke or rod length necessitates decreasing the compression height of a piston for a given block height.

COMPRESSION HEIGHT

Compression height is merely an engineering term for the distance between the wrist pin centerline and the deck surface of the piston. Rod length and stroke combine to establish this distance. With a production small block deck height (measured from the crankshaft centerline) of 9.025 inches, adding half of the stroke and the length of the rod and subtracting them from the deck height will establish the proper zero deck compression height for your piston. Working this out on paper a couple of times is good experience to solidify compression height in your head. Many piston manufacturers do the math for you, indicating in their catalogs the specific rod and stroke combination that matches up with each piston. This is useful information if you decide to start playing around with offset ground crankshafts and longer rods.

PISTON RINGS

The result of using a longer stroke and/or rod is that it moves the wrist pin closer to the ring package. Many companies build pistons where the top radius of the wrist pin bore just barely clears the bottom ring land of the oil ring. While "tight" compression height pistons are the norm in short duration drag racing pistons, they also reduce the thickness of the ring lands, which can present problems. Be sure there is sufficient thickness behind the ring land to support thermal transfer from the ring land and ring. Especially in nitrous applications, the added heat in the chamber can lead to piston ring land failures. If you are experimenting with long stroke and/or long rod motors, it's best to consult with the piston company about the limitations

Notice the difference in placement of the top ring groove between these two pistons. The piston on the left has placed the top ring groove .320 inch from the piston top while the piston on the right is only .180 inch. There are some small performance gains to be made by moving the ring land closer to the piston top. Ring durability improves (especially with nitrous) when the ring is moved further from the piston top.

of compression height for your application. Part of this equation is how close you want to run the top ring from the top of the piston.

Ring Placement

If you inspect a stock or replacement piston closely, you'll notice that the top ring land is positioned quite a distance down from the piston top. It didn't take racers long to start pushing the envelope by reducing that distance. Obviously, ring placement is predetermined by the piston manufacturer for off-the-shelf pistons, but researching the different piston companies might reveal a particular piston that offers an advantageous ring placement. I like to use JE pistons for many of my engine combinations and the top ring is usually placed .250 inch down from the top of the piston. For endurance race engines, some builders tighten this clearance to .190 to .200 inch. Many of the piston companies will specify this distance even tighter to .175 inch as with Manley's custom flat-top small-block piston.

This isn't the tightest that you can run the top ring, however. Many Super Stock and Competition Eliminator racers will run the top ring up to roughly .100 inch from the top of the piston. If you saw one of these pistons, you'd swear you could see the ring right through the top ring land—it's that thin! For short duration drag racing where engine life between rebuilds is measured in minutes instead of miles, the risks are acceptable. For a street engine, my spec of .250 inch is conservative for good reason.

A number of factors determine where the ring package is placed on the piston. Nitrous, supercharging or turbocharging all dramatically raise cylinder temperatures which requires moving the top ring away from the top of the piston to insulate the top ring and to prevent upper ring land failure due to those excessive temperatures.

Ring Sealing

Your efforts to make power will not be worth much if you cannot seal the combustion pressure to maintain pressure on top of the piston. If much of the cylinder pressure ends up in the oil pan, the engine will be a stone. The job of sealing cylinder pressure falls to the piston rings and there are a ton of choices. These choices include material, thickness and ring style.

As you may know, street engines and most competition small-blocks employ a three-ring piston. The top ring is a compression ring, the second ring shares responsibility between compression sealing and oil control while the bottom ring package's main task is oil control. While the top and bottom ring functions are obvious, the second compression ring's main duty is actually to scrape the remaining oil from the cylinder wall while its secondary function is to seal pressure that has leaked past the top ring.

Ring Materials

While all rings may look the same, there are many different materials. Cast iron is a popular OEM ring material. It is inexpensive and works well in production engines, but suffers when used in a high performance application. Since cast iron is somewhat brittle, it is susceptible to failure under detonation. Ductile, or nodular, iron is a far better choice for a top compression ring. The name itself lends the cue since ductile iron is much

The popularity of 1/16-inch rings for street use is increasing since there are small power gains to be made with these thinner rings. This photo compares 1/16-inch (.062-inch) rings (left) to 5/64-inch (.078-inch) rings (right). The thinner rings reduce frictional drag at higher rpm levels. The replacement piston (right) can be identified by the oil return slots cut in the oil ring groove. Pistons requiring greater piston-to-wall clearances usually have oil holes (left) cut in the oil ring groove.

RING GAP

Piston ring end gap is an important specification when assembling an engine. Speed-Pro has produced the following ring end gap recommendations for forged aluminum pistons. Consider these as suggestions and always refer to the specific ring manufacturer when custom-building a combination. Factors that affect ring end gap are the engine's intended usage, the fuel to be used, type of ring, position of the top ring relative to the top of the piston and many other variables. Most ring manufacturers offer .005-inch oversized rings that allow the engine builder to accurately set the ring end gap. The following end gaps are based on a 4-inch bore size.

DUCTILE IRON TOP RING
5/64, 1/16, .043-inch

ENGINE STYLE	END GAP
Carbureted Gasoline Engines	
Street	.016 - .018
Super Stock	.018 - .020
Oval Track	.018 - .020
Supercharged Gasoline	.022 - .024

REGULAR IRON SECOND RING

ENGINE STYLE	END GAP
Carbureted Gasoline Engines	
Street	.010 - .012
Super Stock	.010 - .012
Oval Track	.012 - .014
Supercharged Gasoline	.012 - .014

more flexible to handle the increased loads that can occur with a performance engine, especially with detonation. Since nodular iron is more expensive, often a ring set will feature a nodular iron top ring with a cast iron second ring and stainless steel oil rings.

To improve the sealing characteristics of nodular or cast iron, many slightly more expensive ring sets feature a molybdenum facing, commonly referred to as a "moly" ring. A double moly ring set refers to the fact that both the top and second rings feature moly facings. The more common and less expensive ring sets offer single moly packages where only the top ring is moly faced.

The advantage of moly is its superior seating that virtually guarantees a short break-in time which means the rings "seat" or seal very quickly. Moly faced rings also offer significant improvements over an iron facing in general sealing characteristics and tend to better weather the adverse effects of detonation. All of these reasons point to moly-faced rings as the current most popular choice for both street and race engine.

Chrome-faced rings offer a much slicker surface which would appear to be an advantage. Chrome rings are merely

Most performance ring companies sell .005-inch oversize rings that allow the engine builder to customize the ring end gaps. The best tool for this job is a ring filer such as this Childs & Albert tool. There are also trick electric ring filers if you're willing to spend extra money to save time.

ductile iron rings electroplated with chromium. However, chrome requires special honing preparation that is quite different from either iron or moly-faced rings because of the chrome's greater hardness. While chrome does offer durability advantages, the problems of proper break-in combined with how easy moly rings are to use have virtually eliminated the chrome ring as a choice in all but the most adverse dirt track engine applications. Even then, with the proper air filters, moly rings are still a wise choice.

Ring Thickness

The line "You can never be too rich or too thin" is about as ridiculous for aspiring fat cats as it is for hot rodders. For years, production compression rings in the small-block Chevy have been 5/64-inch. It wasn't long after the small-block became the darling of the race set that engine builders began experimenting with thinner ring sets in favor of additional power. This quest for thinner

rings has led to some exotic designs to take advantage of reduced frictional losses attributed to ring drag. In the search for more power, most Competition Eliminator drag race engines have

eliminated the second ring altogether, assuming that the reduced ring drag more than compensates for the power loss that occurs from oil dilution when the second ring isn't around to control oil.

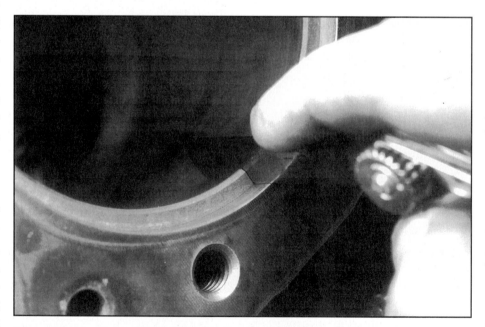

Ring end gap is also crucial to ultimate performance. Too little end gap will allow the ends to butt, which usually leads to a seized ring and a damaged cylinder. Loose end gaps contribute to excessive blowby, crankcase pressure and oil consumption problems. End gaps should be checked with a torque plate installed for ultimate accuracy.

There are wives' tales about how a ring can be damaged by spiraling it on or off a piston, but this isn't true. In fact, ring manufacturers claim more rings are damaged by using ring expanding tools than by spiraling them into place. When spiraling a ring onto the piston, make sure to avoid scratching the piston with the inside edge of the ring end.

Ring thickness has progressed from the stock 5/64-inch (.078-inch) to 1/16-inch (.0625-inch), .043-inch and down to Twiggy-sized .031-inch rings as the more popular sizes. The 1/16-inch rings have been successfully used on the street and even some production engines have recently been fitted with these thinner rings. Rings thinner than 1/16-inch, including the Dykes ring, offer no advantage for street engines.

While at first it may appear that thinner rings reduce friction because of the thinner face contact with the cylinder wall, the reality is a little more interesting. In order to maintain ring seal against the cylinder bore, each ring must have a certain amount of radial tension. As ring thickness decreases, the radial tension is also reduced to maintain the same unit load pressure on the face of the ring. This reduction in unit loading pressure reduces the drag which improves horsepower. This isn't worth a bunch of power, but at 6000 rpm some piston manufacturers will admit to increases of around 10 horsepower. Typical ring combinations for a small-block for a 1/16-inch ring set would be 1/16-, 1/16- and 3/16-inch for the oil ring.

Tension—Production oil rings are typically 3/16 inch thick. Most high performance ring manufacturers also offer oil rings in at least two configurations as standard tension or low tension. The reason for this is because the oil ring is generally considered to contribute the most ring drag. While reducing the tension also reduces drag and power loss, it also generally hurts the oil ring's ability to scrape oil off the cylinder wall. This results in an obvious increase in oil consumption.

Many ring manufacturers such as Speed-Pro, TRW and Childs & Albert offer three different tension level oil rings. Standard tension rings are what would be used in a typical street engine. Speed-Pro defines their standard tension oil rings as having 19-22 pounds of tangential tension. Low tension Speed-Pro rings offer a tension of 15-18 pounds and the Special Light Tension rings produce only 5-10 pounds of tension. These Light Tension rings should only be used in drag race engines using vacuum oil control systems such as the Pan-E-Vac system. Again, small increases in power

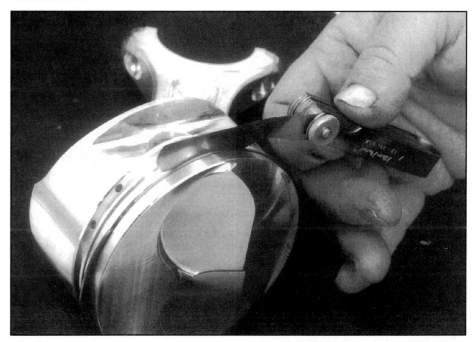

Checking ring side gap is usually overlooked when blueprinting an engine. Most ring manufacturers specify .002 to .004-inch of clearance. Most racers like to reduce this figure to improve control over ring flutter at high rpm. Consult your ring manufacturer for minimum side clearance recommendations.

TOP RING
Positive Twist

Cylinder Pressure

Seal
Seal
Seal

Oil

2nd RING
Reverse Twist

Cylinder Pressure

Seal
Seal
Seal

Oil

The bevel on the inside top edge of the top compression ring creates what is called positive twist where the ring seals on the bottom edge of the ring to prevent combustion pressure from escaping past the ring. Reverse twist rings are usually second rings with the bevel on the inside bottom edge. A reverse twist ring's primary job is to scrape oil off the cylinder wall.

are possible with low tension oil rings, but oil control for a street engine is usually a problem.

Smaller 3mm oil rings are also offered by some ring manufacturers; these equate to an oil ring thickness of .118-inch compared to a standard 3/16-inch ring's thickness of .187-inch. It's possible to reduce friction by using a thinner 3mm oil ring while maintaining standard oil ring tension.

Ring Design

The classic ring design is the D-wall ring which resembles a flat rectangle with a rounded cylinder wall contact face. This is the most widely used ring style for all engines with back cuts placed on the inside diameter of the ring depending upon whether the ring is to be used as a top or second ring. In the quest to reduce ring weight while maintaining a certain ring face thickness, the Dykes ring was developed. The Dykes ring is merely a stepped ring where the ring face may measure .062 inch while the inboard ring thickness is reduced to .031 inch. This

type of ring requires a specially machined stepped ring land. The dimensions of the Dykes or stepped ring may change, but the concept remains the same.

Because of their different tasks, top and second rings are designed differently. The top compression ring is often constructed with "positive twist," featuring a bevel on the upper inside edge of the ring that helps seal combustion pressure from leaking past the ring. The second ring is often designed with a "reverse twist" that places a bevel on the bottom inside edge of the ring. This creates ring twist that helps scrape oil off the cylinder wall. In our "no-free-lunch" world, these "twists" make each of these rings susceptible to leakage of other kinds. For example, the top ring's positive twist can allow oil to leak past while the second ring's reverse twist doesn't particularly seal well against cylinder pressure leakage.

Other design innovations include tapered face second rings that taper the ring from top to bottom to create an oil scraper that improves its ability to remove oil from the cylinder wall.

Gapless Rings—So-called "gapless" rings, or rings using some type of cylinder pressure barrier loss technology have become popular in the last few years. Total Seal and Childs & Albert both offer such rings, and in theory they do offer potential advantages. Horsepower claims are difficult to quantify and I have yet to substantiate these claims to my satisfaction. The Childs & Albert Zero Gap Second ring may offer some torque increases for street engines, especially at rpm levels below peak torque.

It should be obvious by now that careful consideration must be placed on the selection of each and every component that goes into even a mild street small-block. There are no trick parts that will produce a magical 100 horsepower, except maybe nitrous. If you select the right pieces, 10 horsepower here and 10 there will add up in a hurry. That's how the big boys do it—a little at a time. ■

SHORT-BLOCK MACHINING 7

There is an art to building high performance engines that doesn't come out of a box, nor is it something you pour into your engine. It also doesn't come cheap. The art is machine work. Quality machine work isn't even guaranteed with high dollar equipment, although that is often a good indication that the machinist cares about quality. The true art to good machine work comes with the man (𝗼𝗿 𝘄𝗼𝗺𝗮𝗻) behind the machine. We all know the craftsman who can align a race car with a plumb bob and a string and if you're lucky, you know a machinist who can hone a cylinder perfectly "square."

Many hot rodders view machine work as merely a necessary evil on the way to assembling a killer street engine. Knowledgeable engine builders and enthusiasts willingly pay extra for skilled machine work. In this chapter we'll take a look at many of the machine operations that determine the quality of an engine. There are many places where you can pinch pennies by using production rather than aftermarket pieces as a shortcut to save a few dollars. But quality machine work has always been expensive, because quality takes time and good machinists don't have to work cheap to make a living. In fact, reputable shops usually have more work than they can complete in a reasonable amount of time. What this comes down to is the search for excellent machine work is one of the most important steps on the road to a durable,

Although there are some areas where you can skimp to save a few dollars, machine work isn't one of them. Look for a quality machine shop with high performance or racing experience, and a stellar reputation.

high performance street or race engine.

BLOCK MACHINING

The standard machining operations that everybody knows about are boring, honing and installing cam bearings. But there are many more operations that can and often should be done to even basic street engines. Many of these are low skill/labor intensive procedures you can do yourself to save a few dollars. We'll take a look at many of these to give you an idea of the kinds of questions you

should ask when looking for a good machinist. Often, the best place to start is with local racers. If you get a popular opinion from a few racers, chances are you've found your machinist.

Most machine shops use a deck-mounted boring tool to do basic cylinder oversizing. These bars register off of the block's deck surface, so it's critical that the deck surface be flat before boring the cylinders. More accurate boring tools register off the crankshaft centerline which is the best procedure. I use a

Cylinder boring is best performed on a machine like my Rottler boring bar that registers off the crankshaft centerline. This ensures the bore will be perpendicular to the crank throw. This also eliminates the variable of registering off a non-perpendicular deck surface.

Torque plates are these large steel or aluminum plates bolted to the block deck surface that simulate the stress of a cylinder head. It's possible to tell if a block has not been torque plateshoned by installing the torque plates and "kiss" the cylinder wall with a hone. High and low spots will show up on the cylinder wall. I use BHJ torque plates.

Rottler boring machine that does just that. Most shops, including mine, bore the cylinders to within .006 inch of the cylinder bore size and then hone the cylinder to its proper size.

Honing

Cylinder honing is a much debated subject. One thing that everyone agrees on is the cylinder walls must be round.

Here is where racing's fanaticism pays dividends for the street engine builder. Years ago, racers began experimenting with steel plates bolted onto the deck surface of the block to induce cylinder wall stress in the cylinder block similar to bolting on the cylinder head. Fel-Pro has

done extensive research that illustrates the tremendous bore distortion that occurs in the small-block Chevy just as a result of torquing down the cylinder heads. This means that even for a low-budget street engine, spending money for torque plate honing is a good decision.

Honing is as much an art as it is a skill. The popular machine is the Sunnen CK-10, but other variables such as the honing oil used will also affect the work. I participated in early testing of VP's new honing oil that I now use exclusively.

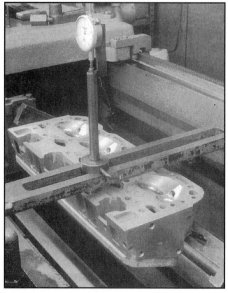

This Kwik-Way horizontal mill is used to trim cylinder block deck surfaces as well as cylinder heads. The operation both ensures the deck surface is flat, but also that it is parallel to the crankshaft centerline.

Rebushing lifter bores will ensure accurate valve timing, but the payoff in more power doesn't justify the significant labor costs for a street engine. Here, the procedure is being performed on a Bow Tie big-block.

Steel four-bolt main caps like these Oliver caps can be fitted to any small-block. The machining operation isn't overly expensive and could be a way to upgrade a block you may already have. The procedure usually requires align-honing the block to register the main bearing bores. Keep in mind that align-honing a current block may mean moving the crank centerline a few thousandths up in the block that will affect deck height and timing chain tension.

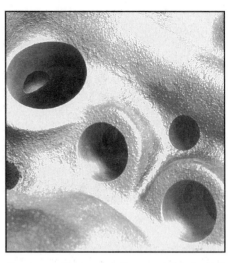

One operation even the most basic hot rodder can perform is deburring a casting by removing casting flash. One area you should concentrate on is the main oil drainback at the rear of the lifter valley. Notice in this photo how this ragged casting flash could break off, enter the lubrication system and tear up the bearings.

The latest torque plates made by BHJ, for example, are stepped to allow proper-length head bolts or studs to be used. Some engine builders even torque the main caps in place to simulate cylinder block distortion. Taking this even further, some machinists claim that if studs are used, they should be used on the torque plate as well, since they change the characteristics of the cylinder distortion compared to head bolts. Most machine shops will also ask you which head gasket you intend to use so that they can duplicate that as well.

For honing, the Sunnen CK-10 has become the standard of the industry. This machine allows the operator to standardize the technique while maintaining accurate control of rotational speed, stone pressure, reciprocation rate and a number of other variables. While there are probably machinists who can produce a very precise cross-hatch pattern with hand-held equipment, it's just smarter to look for a quality shop that uses the Sunnen machine. Then it becomes a matter of the skill of the operator to generate the final finish that will allow the rings to seat properly. Each operator will have his own personal preference in creating this cross-hatch pattern. Unless you have experience with the machine, it might be better to defer to his judgment.

Honing Techniques—There has been much discussion in the industry about "plateau" and "soft" honing techniques. The basic honing process removes the rough machine marks left by the boring bar, smoothing the cylinder wall while still leaving a sufficient cross-hatch pattern in the cylinder wall to help lubricate the rings. Racers prefer a very smooth, extremely round cylinder wall to help ring seal and reduce friction. Some racers have gone to the effort of virtually polishing the cylinder wall in search of that last ounce of reduced ring drag power.

Even after honing, if magnified, the cylinder wall will look like a landscape of peaks and valleys. It's the valleys that retain the oil. Plateau honing is a procedure where the first, rough set of stones is used to hone the cylinder wall. Usually, this is followed by a series of smoother stones that gradually create a smoother surface. In plateau honing, this initial stone is directly followed by a relatively fine grit stone called a cork bond stone. This stone is used to hone off the peaks to produce a significantly different surface. Even though it is smooth, the procedure produces a very rough-looking wall finish. It is this finish that scares many hot rodders from using the procedure.

Some machinists feel the plateau honing process should be used for only very hard ring packages such as TRW's ceramic ring or chrome rings. Others have used a plateau honed cylinder wall with success in street engines. They feel the additional "depth" of the hone offers additional lubrication to the piston and rings. There has yet to be any proven advantages in power to this procedure.

I have experimented with smoother cylinder walls in search of slight power improvements for street engines using moly ring sets. The result of these experiments was no additional power, but it did increase break-in time. With no clear advantage, I have reverted back to standard cylinder wall preparation as outlined in the Speed-Pro catalog.

Decking

Another popular machine operation is ensuring the block's deck surface is both flat and perpendicular to the crankshaft centerline. Often, this operation is used to reduce piston deck height. Production

One of the first operations performed on any critical new or used engine component is to check for cracks. I use a wet Magnafluxing operation that soaks the casting in a luminescent solution that will indicate surface cracks when illuminated with a fluorescent light.

main bearings and then spin the crank by hand. If it won't spin easily or is difficult to spin, then it might be advisable to have the main bores checked. Of course, another possibility is that the crank is bent or twisted, which could contribute to a "tight" crank.

Lifter Bores

You may have also heard about bushing the lifter bores as a way to increase power. This is a popular trick with drag racers in an effort to equalize valve timing for each cylinder. It's not unusual for the lifter bores to be at slightly different angles. This is often determined by different valve timing figures for each cylinder even when the cam is ground accurately. While some believe there are horsepower gains to be uncovered with this exercise, I believe the gains are minimal compared to the cost involved and the accuracy of current cam grinding.

There are a few other minor machining operations such as de-burring the block to remove the casting flash, clearancing the

small-blocks tend to come from the production line with between .020 and .025-inch negative deck height, meaning that at Top Dead Center (TDC) the piston is a certain distance below the block deck surface. Negative deck height both reduces static compression and also hinders the effectiveness of the quench area of the piston-to-deck clearance (see Chapter 6). Decking the block merely reduces the thickness of the deck surface of the block from the production figure of 9.025 inches.

Align-Boring & Honing

Line boring and honing are procedures that are sometimes necessary in street engines, especially if the engine has been pushed to its limits. Most often, line boring is required when fitting the engine with aftermarket steel four-bolt main caps. The procedure involves measuring the inside diameter (i.d.) of the main journal bore. If the bore is too large, the bottoms of the caps are trimmed until the i.d. is roughly the size required. Then a long align-boring tool is used to bore each housing to within a few thousandths

of the i.d. spec. Then an align-honing tool is used to finish hone the main bores to the required spec. For most street engine builders, a quick align-honing check is to torque the crank in place with all the

Installing cam bearings is a delicate job best left to a competent machine shop. The bearings actually come in three different sizes depending upon their position. It's essential that the oil holes line up properly to ensure proper oiling system performance. As you can see, a long driver is needed to install the bearings in the block.

Balancing the crank is important to ensure long bearing life and to limit power-robbing vibration. This is my Hines Industries electronic spin balancer. Weight is bolted to the rod journals to duplicate the big-end's rod and bearing weight. The machine's accuracy is within 1/4 of a gram. I usually shoot for around 1 gram accuracy which is tight considering a typical weight factor of 1900 grams!

block for stroker cranks, installing cam bearings, and fitting the deck surface for coolant restrictors that we'll cover in photos and captions.

CRANKSHAFT

The most popular crankshaft machine operation is turning the main and rod journals on used cranks down to .010-inch undersize to create an optimal round journal. Most engine shops don't do this machine work themselves since a crank turning machine is expensive, so this work should be outsourced to a quality machine shop.

Offset Grinding & Indexing

Offset grinding is a procedure explained in Chapter 4 on crankshafts, however, realize it offers the opportunity to increase or decrease the stroke. Indexing involves precisely measuring the stroke of the crank and optimizing the throw so that all four crank throws are within .001 inch. This is a Stock and Super Stock Eliminator drag racer trick in which the racers take advantage of

NHRA's spec that allows the stroke to be .015 inch more than the published stroke. Most racers prefer to set the stroke at + .013 inch as a safety margin. On a 3.48-inch stroke 350, this is only worth 1.3 cubic inches, but racers will take every advantage possible. Typical cost for indexing is roughly around $100 for a small-block crank.

Balancing

Balancing is another important operation that's important for any street engine builder. Even with replacement pistons and stock rods, balancing removes any doubt about whether all the reciprocating and rotating pieces are compatible. My Hines balancer is like most electronic balancers where the operator bolts bob weights on the crank rod pins that are carefully weighed to duplicate the total weight of the rotating weight of the two rod big ends with their rod bearings. The reciprocating weight is the total of the piston, wrist pin, rings, and the small end of the connecting rod. Simply put, the weight of the crankshaft counterweights should be equal to the

weight of the piston/ring/pin/rod of the reciprocating package. Most crank packages also include the flywheel/flexplate and the harmonic balancer in this balancing operation. Externally balanced engines such as the 400 small-block must include the balancer and flywheel/flexplate since these components are used to set the balance.

One of the components used in calculating the required weight is rpm. The amount of force generated by the rotating components will change depending upon rpm. Some engine builders will "overbalance" the rotating components by one or two percent in an effort to smooth out potential high rpm imbalance. This also requires that the engine builder supply a peak operating rpm to the balance operator so that he can make the appropriate decisions as to how the engine should be balanced. Many aftermarket stroker cranks that increase stroke to and beyond the stock 400's 3.75-inch length can be built as internally balanced engines. If you're interested in being sneaky by hiding the length of your crank's stroke, these engines will not exhibit the classic giveaway of the offset-weighted harmonic balancer of a 3.75-inch cast crank stroke 400.

Cross-Drilling

Cross-drilling is still a popular crankshaft modification that creates a second main journal bearing oil hole in the main journal of the crankshaft. But I believe that this modification does little to improve lubrication and only adds to the machining bill. Chamfering all the trailing edges of the oil holes on both rod and main journals will tend to improve lubrication, but cross-drilling is something that can be left off your machining wish list. A final micro-polishing of the main and rod journals can also be helpful, especially on a used crank that is still within spec or a crank in an engine that is being freshened up with

New rod bolts are inexpensive insurance against catastrophic engine failure. I use ARP rod bolts to keep the big end together even at high rpm. Note the foreground rod has been clearanced slightly for a 383 application to clear the cam.

new rings and bearings. The polishing removes any burrs or sharp edges that could scratch new bearings.

ROD REHAB

While you might think that since connecting rods have no moving parts, why would you have to rebuild it? While the rod doesn't offer a wear surface (assuming the rod bearing does its job), rpm and miles tend to elongate the big end of the connecting rod. Since a round big end is important to extract the most out of the rod bearings, this usually means rebuilding the rods anytime the engine is rebuilt or even when it needs freshening.

If you don't know the history of the rod you intend to use, the best place to start is to have the rods Magnafluxed to check for cracks. The next step is to add new connecting rod bolts, as discussed in Chapter 5. Installing these bolts is best left to a machine shop that can press them in place. You should never hammer new bolts into place. The machine shop will ensure that the bolt head area is flat and chamfered to match the radius in the bolt between the bolt shank and the head. A careful machinist will also make sure no sheared metal appears under the bolt head

that could cock the bolt.

With the rod bolts in place, most machine shops mill a small amount of material from the rod cap to create a smaller i.d. Then the cap is torqued in place and then the big end is carefully honed to bring it back to the standard i.d. spec. I have found that high volume machine shops use courser stones that create heat in the rod big end. Even though they hone the rod to the proper i.d., the big end shrinks as it cools, reducing overall bearing clearance. I recheck all refurbished rods for proper i.d. after the machining operation to ensure proper big end specs.

Careful measurement is also important with big ends since even though the rod may look right, measuring may reveal the i.d. is too small, too big, tapered or bell-mouthed. Any of these situations will reduce bearing life. In worst-case situations, the engine will destroy the rod bearing within minutes of initial start-up. That's where the money you saved from cheap machine work ends up costing much more later on.

If you opt to convert a set of stock rods to full floating pins, the rods are fitted with special high strength bushings. This is also a good time to have the machinist optimize the rod length among all the rods. Unfortunately, this is costly. It's far better to spend a little more money and purchase a set of aftermarket 4340 steel rods that will be much stronger and within a few thousandths of the proper center-to-center length.

By design, this has been a rather broad stroke when it comes to the art of machining operations relative to the street small-block. The main key is to not scrimp when it comes time to spend money for machine work and to not get in a hurry. Spend the time to double-check everything and the money you spend will pay off in an especially durable and powerful small-block. ■

Rebuilding the big end of the connecting rod is usually performed on this Sunnen honing tool. A good operator will perform three or four passes on the hone and then will flip the rod over and perform the same number of passes. This prevents bell-mouthing the rod big end.

ONE PIECE OR TWO?

It won't happen overnight, but the classic small-block two-piece rear main seal engines will soon be replaced by the later model '86 and later one-piece rear main seal cylinder blocks. It is becoming increasingly popular to use a late model, two-piece seal crankshaft in these later blocks, especially if you are using the stronger Bow Tie blocks, which are now machined to accept the one-piece rear main seal. If you are using a two-piece seal crank in a one-piece seal block, this will require the use of a two-piece rear main seal adapter such as the units sold by Diamond Racing and Moroso.

According to the *Chevrolet Power* book (also available from HPBooks), the hollow dowel pins that locate the aluminum seal adapter to the block are not "precision locating devices" that could allow the seal adapter to move that will create leaks. Chevy's recommendation is to use an arbor that accurately locates the seal adapter (without the main seal in place) torqued in place of the crank.

With the seal adapter precisely located, Chevy recommends drilling two interference-fit dowel pin holes through the seal adapter and into the block. With the seal adapter and arbor removed the aluminum seal adapter is reamed to fit the dowel pins. The pins are then tapped into place, acting as precision locating devices to produce a leak-free seal.

A much quicker and simpler procedure is to merely grind off the stock hollow dowel pins that locate the seal adapter to the block. My engine building team has found that these pins are out of position and pull the adapter out of place. By using the stepped arbor to accurately locate the seal adapter to the block, this will prevent further leakage. Either way you go, once the adapter is located properly, it should not be moved even during a rebuild to ensure the seal is located properly.

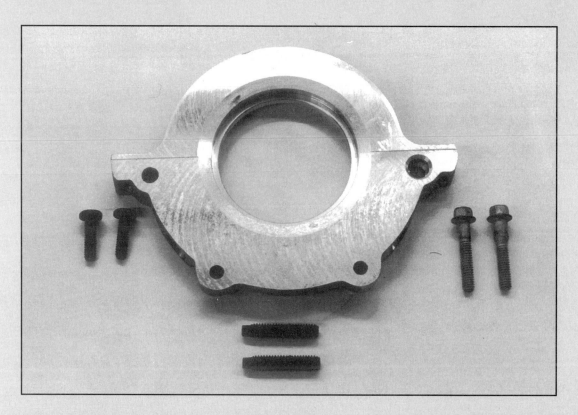

This is the aluminum two-piece rear main seal adapter bolted to the back of a one-piece seal style Bow Tie block. The adapter is bolted to the back of the block with four Allen head bolts. The two lower bolts fit inside hollow dowels that can allow the seal adapter to move slightly.

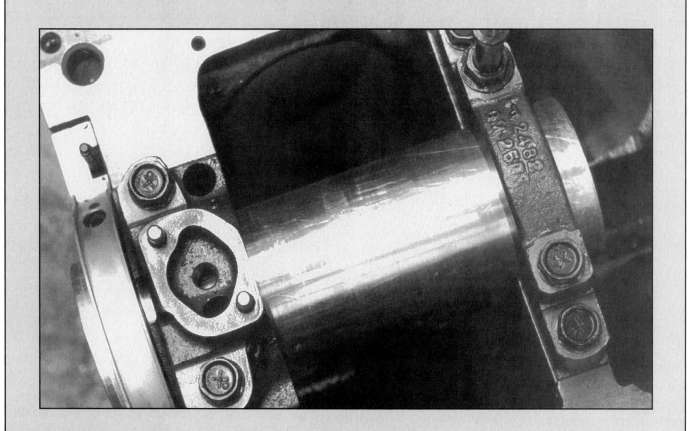

This stepped arbor or gauge plug is used to position the seal adapter in the back of the block. This can be machined out of aluminum to the dimensions shown in the illustration below.

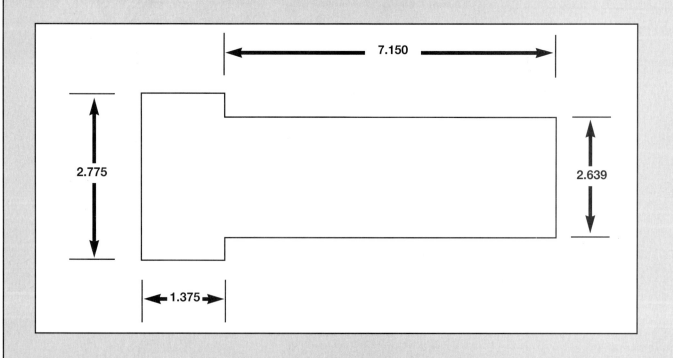

With the block placed in a drill press or vertical mill, drill two small holes through both the seal adapter and into the block to locate the dowel pins. The holes can then be reamed to fit the proper size dowel pins.

These two dowel pins will prevent the seal adapter from moving in relation to the crankshaft and will prevent rear main seal leaks.

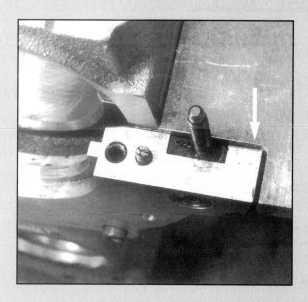

One other trick is to chamfer the right angle edges of the seal adapter where they come in contact with the radius corner of the block. This will prevent the seal adapter from "lifting" by not coming in contact with the block when the adapter is bolted down. One way to check to see if the rear main seal leaks is to fill the pan with oil and hang the engine at a severe tail-down angle from an engine hoist. Leave the engine hanging overnight and if there's no oil on the floor in the morning, the engine probably won't leak in the car.

Offsetting the parting line of the seal so that it doesn't line up with the parting line of the adapter is another trick you can use to prevent rear main seal leaks. The two outboard vertical studs are the rear pan bolts for this two-piece seal adapter in a one-piece seal block.

8 THE LUBRICATION SYSTEM

Perhaps it is a testament to the stock small-block Chevy's oiling system that it requires few modifications to adequately maintain even 500 horsepower street engines. Much of what we will cover in this chapter falls under two categories. The first are merely mild tune-ups to the stock lubrication system to help it do a better job. The second is probably more important since many well-meaning oiling modifications actually hurt performance and durability rather than help. It's important to know the consequences of your actions before you make changes.

As you will see, there is much more to the oiling system than just a pan, pump and filter. Bearing clearances, oil galleys, windage trays, restrictors and even breathers play a part in a well-designed oiling system. Pressure and volume are also important considerations, but not in the way that you might think. Controlling the oil, ensuring it gets where it needs to go and then returning it to the pan to be recycled again is the key to a well-designed oiling system.

THE OIL CIRCUIT

Can you trace the entire oil circuit of a small-block Chevy without looking at the block? While it's not exactly required knowledge to impress the troops at the local Saturday night cruise, it never hurts to know where all that oil goes and how it gets there.

The stock Chevy oil system is so well designed that it can perform quite adequately for engines up to 500 horsepower without major modifications.

We start in the oil pan where the pump pulls oil up through the pickup and into the pump. Pressurized oil leaves the pump and into the rear main cap, making a right turn into the oil filter boss. The oil travels down the outside of the oil filter, through the filter media and up through the center of the oil filter into the filter adapter. The oil then travels up an angle-drilled oil passage in the bellhousing area and then back down into main oil galley that runs down the center of the block atop the camshaft journals. The next time you happen upon a bare block with the cam bearings removed, look for the recessed grooves located behind each of five cam bearings. These grooves channel oil around the outside of the cam bearings and then to each main bearing via a drilled passage. Each main bearing has holes that pass the oil to the crank. The crankshaft is internally drilled to channel oil from the mains to the rod journals. Oil then leaks past both the main and rod journals, returning to the oil pan to start the process all over again.

The main oil galley in the lifter valley also feeds two parallel oil galleys that are

64

A low buck trick that does work to increase oil pressure with a stock spring is to install one or two small AN washers under the spring to increase pressure. Be conservative here and keep the washers limited to two. Be sure that adding washers does not prevent the relief valve from opening fully.

There are very few tricks needed to make a small-block Chevy oiling system perform. Chevy's classic Z/28 oil pan, windage tray (top), tray studs and the standard oil pump with the white Z/28 pressure relief spring is all you really need.

drilled through the middle of the lifter bores. These galleys feed oil to the lifters, which then direct oil up hollow pushrods to squirt oil onto the rocker arms and valve springs. The oil then returns back to the pan by draining from the heads to the lifter valley where a number of oil drainback holes to the pan are located. Oil is also directed past the front cam lobe to the fuel pump pushrod as well as to the timing chain. The distributor is lubricated via the left side oil galley.

OIL PUMPS

The bottom line for a small-block Chevy street engine's lubrication system comes under the heading of "If it works, don't fix it." The stock factory system works extremely well with a stock pump, stock pressure and stock volume. Many companies sell high performance pumps that will increase the pressure, the volume or both. Most of the time, added pressure or volume is unnecessary. This is not to say you can get by with the least expensive pump out there that barely does the job. Instead, the smart move is probably a well-built new pump from Sealed-Power, TRW, Mellings or others and fit it with a mild performance pressure relief spring and a brazed pickup and you're more than halfway there.

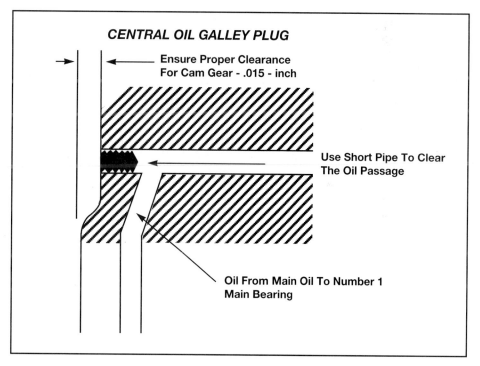

CENTRAL OIL GALLEY PLUG

Ensure Proper Clearance For Cam Gear - .015 - inch

Use Short Pipe To Clear The Oil Passage

Oil From Main Oil To Number 1 Main Bearing

Most performance engine builders tap the three front oil galley plugs for screw-in plugs that replace the pressed-in cup plugs. The middle plug should be shallow tapped and you should use a short-depth plug to prevent overlapping the vertical oil passage that feeds the Number One main journal. If the oil plug restricts this passage, this will reduce oil flow to the Number One main that could damage the bearings.

Stock Pumps

Stock Chevy oil pumps create pressure by squeezing oil between a pair of gears driven by the distributor shaft. Maximum pressure is limited by the pressure relief spring. The spring is enclosed in a chamber housing a small piston. When the pump pressure exceeds the spring pressure, the piston unseats and bleeds off a small amount of oil to maintain the pressure. The oil is merely recirculated back through the pump rather than

Oil temperature is an important factor in performance. Often, even if engine coolant temperature is within reason, the oil temperature can still be too hot or too cold. There is some debate over minimum oil temperature for performance applications, but it should be at least around 210 degrees F. Optimal oil temperature is 220 degrees F. while anything over 250 degrees is dangerous ground for petroleum-based oils. Synthetic oils can withstand higher temperatures in the 250 to 260 degree F range for sustained periods. The only way to know oil temperature is to measure it with a gauge. This small fitting above the oil filter boss is a great location for an electric oil temperature probe.

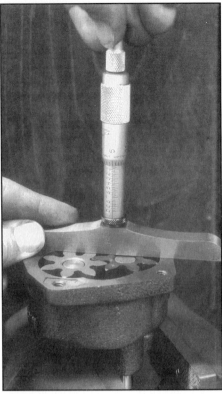

Blueprinting the oil pump means checking the clearance between the oil pump cover and the gears. Excess clearance reduces the pump's efficiency. The clearance can be checked by using a depth micrometer to measure the distance the gears are down from pump body. Recommended clearance is between .0025 and .0033 inch. If the gears are more than .003 inch below the pump body, sand the pump housing on a flat surface until the spec is reached. If the clearance is less than .0025 inch, sand the tops of the gears.

Stock oil pump shafts will work fine in most street applications. Perhaps the only weak point is the plastic collar (left) that connects the driveshaft to the oil pump shaft. Performance shafts use a steel collar (middle) that is more durable. The 400's larger main journal diameter requires a necked down oil pump drive shaft (right) to clear the larger journal. If a standard shaft is used, it will rub on the block and eventually fail.

dumped back into the pan. This tends to heat the oil, which is not ideal, but works fine for the street.

Many aftermarket companies offer high pressure, high volume pumps that combine the two functions. But given the general 10 psi per 1000 rpm rule for oil pressure, there's little need for street engines to require more than a 70 psi oil pump. This is easily attainable with a stock volume pump fitted with nothing more than Chevy's white, Z/28 oil pressure relief valve spring. This compares to roughly 50 to 60 psi for pure stock small-block pumps. If you are conscientious, it's a good idea to disassemble the pump to check the clearances. A depth micrometer check of the clearance between the oil pump cover

and gears should produce a spec of between .0025 and .0033 inch. If your pump is equipped with a gasket between the cover and the gears, be wary. Only low quality pumps come this way. Drop kick that rascal into the trash can and buy a better pump.

Pressure and Volume

Enthusiasts and budding engine builders often confuse pressure and volume. While these two variables are related, an increase in one does not mean an increase in the other. Oil pressure is created when the two gears in the pump pull oil through the pickup and into the cavity of the oil pump. As the oil travels past the gears, it is squeezed into a smaller space. Unlike air, oil is non-

It never hurts to radius the point where the oil exits the pump and enters the rear main cap and also around this area where oil enters the block from the filter. This is a simple procedure and can be accomplished with a die grinder and a couple of hard rolls.

All three of these oil filters will fit the post '67 small-block Chevy. The short filter can be used in tight header configurations, but if you have the room, the taller filter with more filter area is preferable. The Fram HP4 is designed for high pressure applications. While more expensive, it also creates less restriction to oil flow through the filter. For most street engines, the Fram is not necessary.

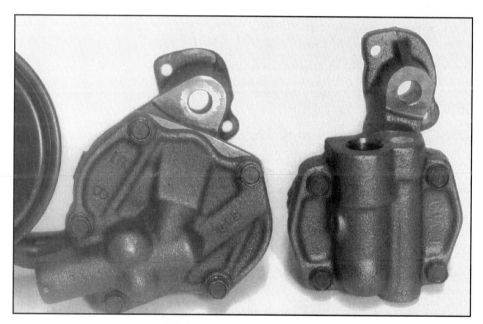

The easiest way to tell the difference in small and big-block pumps is the number of cover bolts. The small-block pump (right) has four bolts while the Rat motor pump (left) has five.

the pump. The other is nothing more than a big-block Chevy pump which is already fitted with larger gears. Both of these pumps increase volume over the stock small-block pump. In either case, pressure is increased by changing to a stiffer pressure relief spring in the pump or by shimming the spring.

There are very few good reasons for using a high volume pump for the small-block since the stock small-block pump is fully capable of supplying the needs of even the heartiest street engine. In fact, a stock pump with no oil galley restrictors is capable of pumping perhaps as much as half of the engine's total oil supply into the valve covers at sustained high rpm. Given this capability, you might see why there is little need for a high volume oil pump.

New car manufacturers have been steadily reducing oil pressure in production engines in an attempt to gain some slight advantage in fuel economy. This is because creating oil pressure requires power. Try spinning an oil pump with a preluber and cold engine oil using a 3/8-inch electric drill motor and you will see that even a stock oil pump requires a fair amount of power. Multiply that by the peak rpm and power lost to the oil pump is significant. That's why oil pressure beyond the 10 psi/1000 rpm rule will only cost horsepower. For street engines, excessive oil pressure will also lower fuel mileage. High pressure and high volume also place a much greater load on the distributor drive gear.

Big Block Pumps

You may have read in other publications about using a big-block pump in a small-block. The Rat motor pump is identified by the five bolts used to retain the pump cover. This pump uses larger gears with a larger pump inlet opening for more volume. I don't recommend it, and see no reason to use any high volume pump for the small-block. The larger pump merely creates

compressible so the act of shoving the fluid through a smaller space reduces the volume at the same time that it creates pressure. In order to get oil throughout the engine, you must have pressure. Too little pressure and the oil may not reach throughout the engine, especially at high engine speeds. Too much pressure is detrimental because it requires more engine power to create. The general rule that works for the small-block Chevy and

other engines is 10 psi of oil pressure per thousand rpm. Therefore, a street engine that may see 6000 rpm would need 60 psi of oil pressure at maximum engine speed. Idle pressures usually hover around 20 to 30 psi at normal operating oil temperature.

There are two kinds of high volume oil pumps for the small-block Chevy. One uses .250-inch taller gears in a taller pump housing to increase the volume of

Before welding the pickup to the oil pump, it's best to mock up the pump assembly to check the clearance between the pan and the pickup. The easiest way to do this is to place a lump of clay on the flat portion of the oil pump pickup. Then bolt the oil pan with a couple of bolts. The thickness of the clay is the distance between the pickup and the pan. The recommended distance is between 3/8 and 1/2 inch. Don't forget to install the oil pan gasket when checking the clearance.

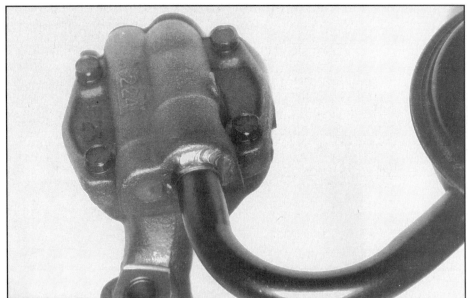

Once the oil pump pickup has been placed in the cover for the proper pan clearance, it can be brazed or welded to the cover. Make sure to remove the oil pressure relief spring from the cover before welding. This will prevent damaging the spring.

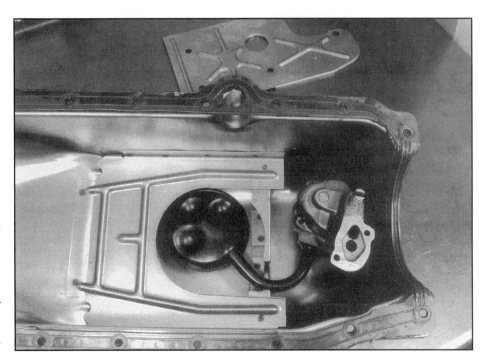

The late model L-98 aluminum head Corvette 350 engine uses a great baffled oil pan that helps keep the oil centered around the pickup. This is far superior to a bare oil pan with no baffles or windage trays of any kind.

more problems while offering no benefits and requires more power to turn! The necessity of using a large volume oil pump is usually an indication of lubrication system problem elsewhere that, once repaired, negates the requirement of a high volume pump. High volume oil pumps, especially if combined with greater pressure, also place a greater load on the distributor gear. If you are using a bronze type gear with a steel roller cam, this additional load will radically decrease the distributor gear longevity.

Another way to identify a high volume small-block pump from a standard pump is to measure the height of the oil pump gears. Stock small-block pump gears measure 1.200 inch while high volume pump gears are longer, measuring 1.500 inch.

Welding Pickups

Everyone recommends brazing or welding the pickup to the pump, which is excellent advice. The stock pickup is merely press-fit into the pump cover and could vibrate loose. Before welding, press the screen into the pump and measure the distance from the screen to the bottom of the pan. A clearance of between 3/8 and 1/2 inch is preferred. Remove the relief valve spring from the cover before welding the pickup to the cover to prevent heat-damaging the spring. After welding, ensure that the

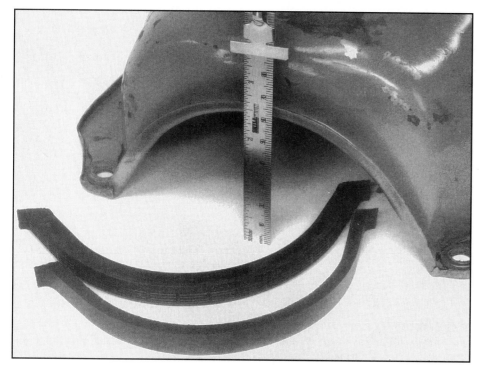

This is an example of a performance oil pan from Hamburger's that places the pickup in a deep sump that keeps oil in the sump with one-way swinging trap doors.

One of those pesky items that trips up many first-time Mouse motor builders is the difference in front oil pan seal thicknesses. In '55 to '74 small-blocks, pans utilized a thin front seal while '75 to '85 engines use the thick seal. One way to tell the difference is to measure the distance from the oil pan rails to the bottom of the front seal lip as shown. Thin seal pans measure 2 1/4 inches while thick seal pans measure 2 3/8 inches. The '86 to current pans also use the thick front seal but with the one-piece rear main seal. This seal choice is easy to overlook since a thin seal used with a thick seal pan will appear to seal properly when, in fact, you'll be blessed with a huge oil leak. Don't miss this!

cover has not warped from excessive heat. This can reduce the efficiency of the pump or even gall the cover or the gears.

WINDAGE TRAYS

Windage trays are another excellent modification for the small-block. Again, there are hundreds of aftermarket units available, but through the years, the most popular is the tested, tried and true Z/28 oil pan and windage tray combination. Chevrolet originally designed the tray and oil pan for the '69 Z/28 engine, but it has found favor with hundreds of thousands of small-blocks.

The tray is designed to nest into a small lip that is incorporated into the matching Z/28 oil pan to direct oil away from the crank and back to the pan. The system also requires five studs that replace the stock main cap bolts to support the tray. While the windage tray and pan were originally designed for the 302 small-

block, the combination will work equally well on all small-blocks, including the 400. Since the 400 uses larger main bearing journals, the holes in the tray must be slightly elongated with a rat-tail file to accommodate the wider-spaced main cap bolts.

Of course, there are literally dozens of aftermarket windage trays, scrapers and trap door-style oil pans that offer increased oil control and they do work very well. Many of these pans offer increased oil volume that keep the oil pump submerged, ensuring adequate oil pressure. This is important if you intend to run the engine at sustained rpm for long periods of time or if you have mistakenly chosen a high volume oil pump. Deep sump pans keep the oil away from the crankshaft, but they do so at the cost of reduced ground clearance. Road race oil pans accomplish this by adding "kickouts" that increase sump volume but are often difficult to use with chassis

headers.

If you run into oil pressure fluctuations on a well-handling car or a drag car that leaves especially hard on the starting line, your first thought may be to add another quart of oil to the engine to maintain sufficient oil level in the pan to prevent the pickup from sucking air. While this may work as a temporary fix, it's best to pull the oil pan and fix the problem. Small-blocks are often built with no windage tray of any kind in the pan that contributes to oil control problems. Usually, the addition of a windage tray will solve the problem. Merely adding another quart of oil will tax the already overworked piston oil rings and will cost horsepower with additional windage and will also increase aeration of the oil.

RESTRICTORS AND BREATHERS

Earlier in this chapter, we traced the entire oiling circuit from the oil pump back to the pan. At the intersection of the angled passage from the oil filter into the two main lifter galleys there is a .250-inch diameter hole. This can be accessed through the two rear main oil plugs inside

The oil filter adapter also houses a bypass valve that allows oil to route around the oil filter. The advantage to plugging this bypass with a pipe plug (left) is that all the oil must go through the oil filter.

If your engine is equipped with roller rockers and a mechanical roller cam, you might consider a set of oil restrictors. These are found in the bellhousing area of the block. Never use restrictors in an engine equipped with stock rockers or hydraulic tappets.

Breathers are an important part of any engine package. Cylinder pressure leakage past the rings tends to pressurize the oil pan and can blow out gaskets if not properly vented. Oil separators in the valve cover also help prevent oil from spilling out onto the valve covers.

the bellhousing area of the block. If you are using hydraulic lifters and/or stock rocker arms, this area should remain stock. However, if you are using solid or roller lifters and/or roller rocker arms, it's possible to incorporate oil passage restrictors that limit the volume of oil directed to the lifters and upper end of the engine. These restrictors tend to keep more oil in the pan and limit the common small-block problem of pumping excess oil to the valve covers. Competition Cams, Moroso and others sell these restrictors and by merely tapping the rear of the block, they are easily installed. The typical restriction reduces this .250-inch passage to around .060-inch for engines using a roller cam and roller rockers. For a flat-tappet solid cam using roller rockers, a larger .090-inch restrictor should be used.

Since stock rocker arms pivot on a ball, they require sufficient oil to limit heat buildup and galling. Hydraulic lifters also require a sufficient amount of oil to operate properly. For these reasons, restricting oil to the upper end for these engines is not recommended.

Blocking Bypass Valve

Another common circle track trick is to block off the oil filter bypass valve. This spring-loaded valve is incorporated into all small-blocks that allows oil to bypass the oil filter should it become plugged through lack of maintenance. This may be great for your grandmother's Malibu, but no self-respecting hot rodder would ever neglect his engine this badly. The disadvantage to this bypass is that its lightweight spring allows unfiltered oil into the engine, especially in cold startup conditions when pressure is high. By tapping the oil filter adapter for a 1/4-inch pipe plug you can ensure that all the oil that travels through the engine will be completely filtered first. This can create rather high initial oil pressures when you first start a cold engine, which could cause a split oil filter if the engine is revved too high on cold startup. Of course, no knowledgeable enthusiast does this anyway since we all know that's rough on pistons and cylinder walls, right?

Breathers

Breathers are another important consideration that should not be overlooked. Since no engine can completely seal all combustion pressure in the chamber, a certain amount of pressure leaks past the rings and occupies the crankcase area. If this pressure is not relieved, it can internally pressurize the engine sufficiently to blow out an oil pan or valve cover gasket. Excess pressure can also push the dipstick up out of the tube! The best way to prevent this is to adequately vent the engine with valve cover breathers. Since excess oil can collect in the valve covers, many valve covers come with convoluted oil separators that prevent oil from climbing out of the valve cover and into the breather where it leaks all over the engine. If you are not restricted by emission requirements, two large K&N

style breathers on top of the valve covers is more than sufficient to ventilate the crankcase. Keep in mind that the better the engine is sealed through proper machine work the less crankcase pressure will escape into the crankcase, reducing the load on the breathers.

One of the best tools you can purchase for your new small-block is an oil preluber. Starting a new engine, even still fresh with engine assembly oil, is dangerous with all the fresh parts. Prelubing the engine drives the oil pump with an electric drill motor to pressurize the entire lubrication system before the engine is started for the first time. This not only ensures that every component in the engine is adequately lubricated, but it also tends to sweep away small particles

of dirt, lint or metal that may still remain in the engine despite your best cleaning efforts. Many engine builders also turn the engine over a few times by hand while pressure lubing to ensure all the bearings receive proper lubrication. The best way to tell the entire engine has received oil is to keep pressurizing until oil reaches all the rocker arms through the pushrods.

As you can see, the small-block Chevy lubrication system borders on the bulletproof right out of the box. If all you add is a Z/28 oil pan, windage tray and a Z/28 pressure relief spring, you've probably got most of what a hot street small-block should need to live a long, healthy life. ■

CHEVROLET PERFORMANCE PARTS

Many of the parts necessary for the recommended mild upgrades you can perform to any small-block are found right at your local Chevy dealerships. While there are many high quality aftermarket parts available, the Chevy parts are proven performers, and as close as your nearest dealer.

COMPONENT	PART NUMBER	QUANTITY
Pressure relief spring, Z/28, 70 psi	3848911	1
Oil pan, '69 Z/28 style	465220	1
Windage tray (for use w/ 465220 pan)	3927136	1
Studs, windage tray (w/ '68-newer blocks)	14087508	5
Studs, windage tray (w/ pre-'68 blocks)	3872718	5
Oil pan, Corvette, 6 qt. ('86-newer)	10055765	1
Oil pan, Corvette, 5 qt. ('86-newer, H.O. 350)	10110837	1
Windage tray (for 10055765 pan)	14071077	1
Oil pan reinforcement ('86-newer, left side)	14088501	1
Oil pan reinforcement ('86-newer, right side)	14088502	1
Oil pump, std. volume w/ Z/28 spring	3848907	1
Oil pump pickup (use w/ 465220 pan)	3855152	1

Note: All pre-'80 small-blocks place the dipstick on the left-hand (driver's) side of the block while '80-'85 blocks moved the dipstick to the right-hand (passenger) side. Both of these are still two-piece rear main seal oil pans. The one-piece rear main seal blocks began in '86 and locate the dipstick on the right-hand side.

9 CAST IRON CYLINDER HEADS

There is no single component on any engine that plays a greater part in creating power than cylinder heads. While it is true that all the other components do play a significant part in this power equation, the heads are the key. I feel so strongly about cylinder heads that if you can spend money on an engine in only one area, the heads are the place to be. Of course matching the other components to the heads is still critical, but you will never go wrong by emphasizing cylinder heads above everything else. To put it another way, an engine with a weak set of heads and the most ideal camshaft will be a stone. But an engine with an excellent set of heads and a marginal camshaft will still make great power.

This is an easy equation since the popularity of the small-block Chevy offers such a tremendous number of heads from which to choose. Remember that this simple good/bad equation is not limited to heads that are too small. A "bad" head can just as easily be one with oversized ports.

Because there are so many different choices and variables, the subject requires two separate chapters. This chapter will cover small-block iron heads, both production and aftermarket, while Chapter 10 will focus on aluminum cylinder heads. There are advantages and disadvantages to both materials and we will attempt to cover all the bases so that you can make an intelligent decision

Production heads are not all created equal. I prefer one of three stock iron 76cc small-block heads. All small-block heads can be identified by the last three digits of the casting number. The best place to find this casting number is in between the valve springs on top of the head. This is a 441 casting.

about which head is best for you.

PRODUCTION HEADS

While the lure of all those trick high performance heads is attractive, the majority of street small-blocks are powered by stock production iron cylinder heads. They work well and will deliver hundreds of thousands of miles of trouble-free performance if properly maintained. While stock castings may not create phenomenal power potential, there are a few that offer good performance potential. Listing all of the production

small-block Chevy cylinder heads and detailing their specifications would easily fill a book this size and, thankfully, many of these castings are now so rare that they would be difficult to find. The reality check for today's mild Mouse motor has limited the useful heads to only five or six castings.

993, 487, 441, 442 Castings

I prefer to use an aluminum performance head like the ported Corvette head whenever possible since it

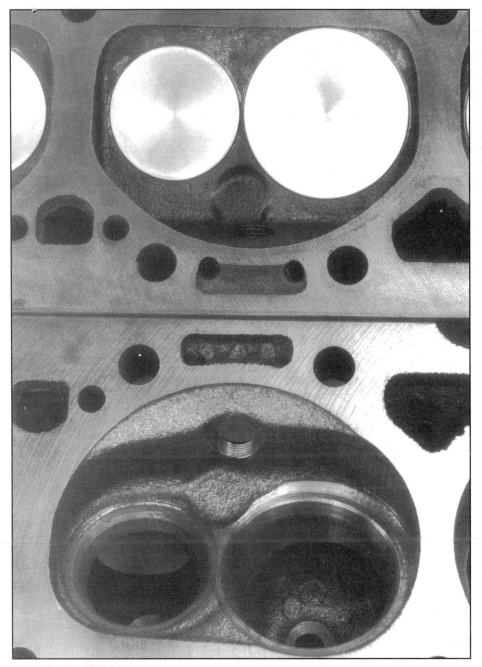

Pre-'69 small-block heads came with 64cc combustion chambers (bottom) while later heads for the 350 engine were outfitted with 76cc combustion chambers (top). A 76cc combustion chamber with a .020-inch piston deck height, flat top 4-valve relief piston and a .038-inch gasket will produce an 8.73:1 compression ratio. Swap on a set of 64cc heads and the compression jumps over a full point to 9.86:1.

Late-model heads with larger chambers all come with accessory bolt holes drilled in both ends of the head. These bolt holes mount the alternator, power steering and A/C brackets required for long water pump applications. Earlier 64cc heads do not have accessory bolt holes drilled. Keep this in mind when swapping water pumps and accessories.

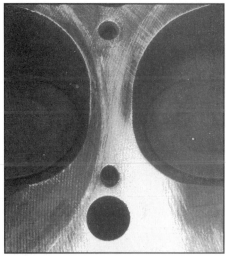

The 400 small-block requires steam holes drilled between the combustion chambers that match up to holes in the block and head gasket. It's important that these holes be drilled in order to prevent low speed overheating problems. These holes relieve steam pockets that build up between the 400's siamesed cylinders.

delivers excellent power and light weight. But when this isn't possible, I prefer one of four iron production heads: 993, 487x, 441 or the replacement 492 casting. These numbers are the last three numerals of the casting number and it is the easiest way to identify small-block cylinder heads. The old days of looking for a set of "double hump" heads are history. Of these four castings, the 441 head is now the best known and therefore is already becoming difficult to find despite the fact that tens of thousands of these heads were produced. The 487x head has perhaps the largest intake port volume but all flow roughly similar on the flow bench.

Chamber Sizes—All four of these heads are the larger, '70s emission-style 76cc combustion chamber size usually fitted with 1.94/1.50-inch intake and exhaust valves. Since the 350 is the most popular small-block, 76cc heads have become particularly important since a flat-top piston with a normal deck height and composition head gasket creates around a 9:1 compression ratio. Reducing the piston deck height will further increase compression up to around 9.5:1 which is the maximum for iron head

Port volume is an easy way to determine the size of an intake port. While a larger port volume does not always guarantee increased power, it is a useful indicator. Stock production heads are typically around 160 cc's while the Bow Tie head measures 184 cc's. CDI offers a great inexpensive kit that will allow you to measure both combustion chamber volume (shown here) and port volume.

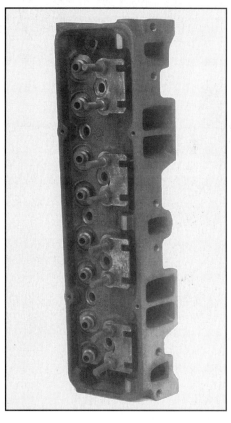

Chevy's two iron replacement heads come with either 64cc or 76cc combustion chambers. Both heads offer flow potential similar to a stock production head. The heads are machined for 2.02/1.60 valves and come with screw-in studs and guideplates.

All three of the 441, 993 and 487x heads flow very close to the same, making them the best choice for a street engine if stock heads are all you can afford.

street engines on 92 octane fuel. Earlier 64cc chamber heads, if you can find them, will pump the compression well over 10:1, which is too much for an iron head engine running 92 octane pump gas with a mild camshaft.

Durability—The 993, 487x and 441 used heads offer not only the most flow potential, but also the most durability. These heads were used in earlier 350 engines before Chevrolet went to the lightweight iron castings that are susceptible to cracking. The 882 heads are easily found in bone yards and at swap meets but not only do they not flow as well as the aforementioned castings, but they are prone to cracks usually in the combustion chamber area. While used heads with small cracks away from a water jacket can be used successfully, it's an unreasonable risk when you consider the investment to rebuild a set of used heads. This is why Magnaflux inspecting is the very first operation before machining a set of used heads.

492 & 624 Castings

Chevy also offers two new replacement iron cylinder heads in both 64cc and 76cc combustion chamber sizes. As you can see from the chart on page 78, the 492 is the replacement casting for all the earlier 64cc chamber heads such as the 461, 462 and 186 heads to name a few. This head comes machined for the larger 2.02/1.60-inch valves along with screw-in studs, and guideplates, although the valves and springs are not included. One great feature is that this replacement head is fitted with accessory bolt holes drilled on both ends of the head, making it great for use with the more popular long water pump accessory systems such as the alternator, power steering and A/C that require these bolt holes. Unfortunately, these heads don't flow as well as those

The iron head on the bottom is outfitted with a heat riser (arrow) which is a passage that detours a slight amount of exhaust gas from a middle exhaust port to heat the area underneath the carburetor to improve cold climate driveability. The aluminum 'Vette head on top eliminates this heat riser passage. Many iron and aluminum heads offer optional blocked heat riser passages for non-emission applications.

Lightweight head castings (bottom) can be identified by the cutouts around the lower row of head bolts. Note that the earlier casting (top) does not have the cutouts. Lightweight iron castings tend to suffer cracking problems after long term use. If you plan to use a lightweight casting, be sure to check for cracks by Magnafluxing.

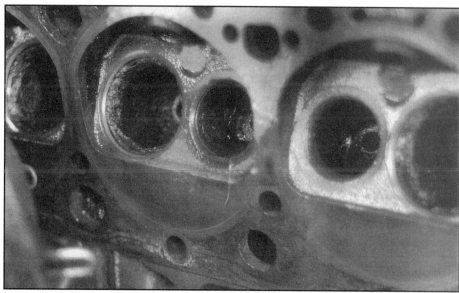

Magnafluxing involves a liquid penetrant dye that can be viewed under a black light. Here, you can see an obvious crack coming from the exhaust valve seat.

earlier 461 and 462 castings.

The 624 casting head is identical except for a larger 76cc combustion chamber. Both heads are outfitted with a heat riser passage for quicker warm-up on cold winter mornings. If there is a down side to these heads, it is that they tend to be overpriced, especially when compared to reconditioned heads from the volume rebuilders. While the Chevy heads are brand-new castings and have not been abused by 100,000 miles of wear and tear, the price rarely justifies whatever small advantage they may promise.

BOW TIE HEADS

As you can see from the Iron Head Guide, most of the production-based small-block heads fall in the 160cc intake port area, which creates decent airflow for low and medium engine speeds, but restrict the airflow necessary to crank out high rpm power. The next step up is Chevy's iron Bow Tie head with its 192cc intake port volume. The Bow Tie head was introduced in 1979 as Chevy's upgrade to the initially successful Turbo iron casting that had a smaller, higher velocity 182cc intake port and a 64cc chamber. Unlike its aluminum cousin, the iron Bow Tie head has remained essentially unchanged since 1979 from an airflow standpoint. Chevy has changed casting techniques, removed the heat riser passage and created a machined pad that runs the length between the exhaust ports. This is the easiest way to identify the later Bow Tie casting.

In the mid-'80s, the iron and aluminum Bow Tie heads were the target that everyone aimed at and were considered

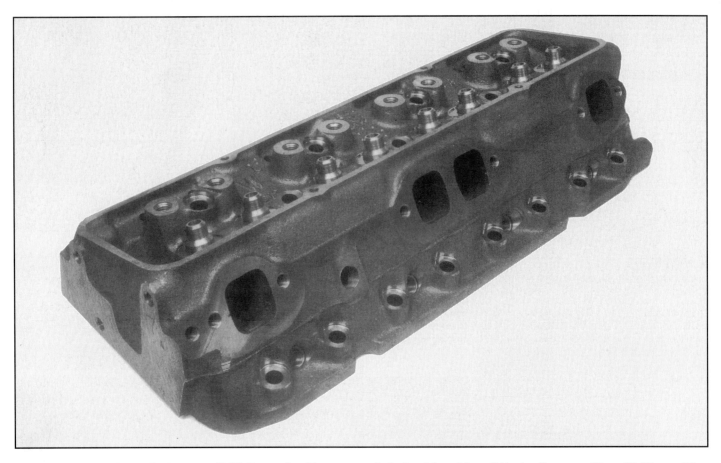

The World Products Street Replacement (S/R) Torquer head is a great cylinder head for mild small-blocks. It comes either with 67cc or 76cc chambers, 2.02/1.60 valves and machined for screw-in studs and guideplates. The head is offered either fully machined but bare or as an assembled head. The price is right too.

an excellent head for both street and racing. Its 192cc intake port is a great compromise between torque and top end power. Unfortunately, the iron Bow Tie head has never enjoyed a competitive price and many aftermarket aluminum heads are now capable of beating both its flow potential and its price while also offering reduced weight. But if you run across a set of used iron Bow Tie heads for a great price, these heads do offer a great step up over a stock set of production heads.

AFTERMARKET IRON

Despite the fact that over 100 million small-block Chevy heads have been produced since 1955, the number of rebuildable cores for the average small-block Chevy continues to dwindle. This is due to both tremendous demand and the fact that most late Chevy lightweight castings crack.

World Products

This led World Products to produce the Street Replacement (S/R) iron head that is a direct replacement for the typical 1.94/1.50-inch valve 76cc early '70s small-block head. There's a 64cc chamber S/R version as well. While significantly stronger with a thicker deck surface, it has a revised exhaust port that improves flow slightly over stock. World Products offers the head in a number of configurations either as bare machined castings or complete heads. The price is competitive to what you would spend to rebuild a set of used castings.

S/R Torquer—World then upgraded the S/R head to larger 2.02/1.60-inch valves and called it the S/R Torquer. The Torquer has the same 171cc intake port and can be ordered either complete or as a bare head. World Products also offers this same S/R Torquer configured with a

smaller 58cc chamber with revised valve centerlines to bolt on a 305. Given the price, the S/R Torquer looks to be a good deal for strict budgets. With a dual pattern cam this head can make 400 lbs-ft. of torque on a pump gas street 350.

Dart II

The step up from the S/R is the Dart II iron performance head. The Dart II Sportsman 200 is a larger version of the iron Bow Tie with a 200cc intake port and is offered with either 64 or 72cc chambers. The Sportsman head is also configured for screw-in studs and guideplates with a thicker deck surface than production heads. This 200cc port volume makes the Sportsman too large for a mild street 350 or even a mild 383, but could work in a stout 383 or 406 cid. Like the S/R head, the Sportsman is available either as a bare, machined casting or as a complete assembled head

Dart Machinery builds a 200cc intake port iron performance head called the Sportsman. The head is offered either bare or complete and ready to bolt on with 2.02/1.60-inch valves, screw-in studs and guideplates. This would be a high rpm horsepower cylinder head but tends to sacrifice torque at street-driven engine speeds.

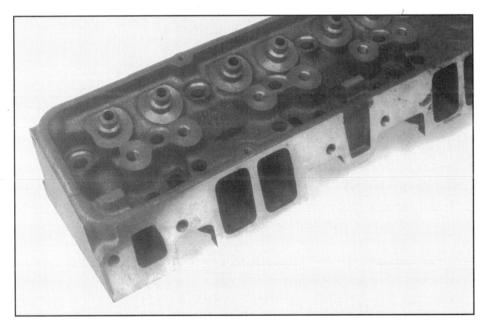

The Chevy Bow Tie has changed little in the last 15 years but still offers great overall power with its 184cc intake port. This is a great compromise size that allows the engine builder great street engine flexibility. The head is fitted with angled spark plugs and no heat riser passage.

Pocket porting involves massaging the pocket or bowl area underneath the valve to reduce the restriction just below the valve seat area. Opening this throat area to 85 or 90 percent of the valve diameter is ideal.

IS BIGGER BETTER?

Many enthusiasts feel that adding larger valves to a pair of production iron heads is a quick way to increase power. Unfortunately, this isn't always the case. My experience has shown that often adding a 2.02/1.60-inch valve combination to an iron head with stock 1.94/1.50-inch valves can actually hurt flow or, at best, not improve air flow. This is due to the restricted area directly underneath the valve seat called the throat area. Testing indicates that intake and exhaust flow will often drop with larger valves at certain valve lifts and that larger valves rarely contribute to better flow numbers all the way through .600-inch lift!

While this isn't the case with every head, it does point out that larger valves don't automatically increase airflow. However, if you open up the bowl area (called pocket porting, see Chapter 11) at the same time that you add the larger valves, that combination is certainly worth the effort. Testing has revealed that this can be worth in the neighborhood of 25 horsepower. Of course, pocket porting can also improve aluminum heads as well. Usually, larger valves tend to increase the length of the port's short side radius which improves low lift flow. ■

with the options of straight or angled spark plugs and the choice of 64 or 72cc combustion chambers.

As you can see, there are plenty of choices in iron heads. Actually, when you get into the iron pieces such as the Bow Tie and the Sportsman heads, there is also the subject of weight and cost. Both of these heads are priced in the area of $1000 for a complete set. For a little more money you could pick up a set of aluminum heads. The aluminum cylinder head market has literally exploded in the last few years with a huge array of choices and options. Plus this additional investment in aluminum also trims anywhere from 40 to 50 pounds off the front end of the car. We'll take a closer look at aluminum heads in Chapter 10.

IRON HEAD GUIDE

The following is a reference guide for both the popular small-block Chevy iron cylinder heads production-based and those available new from Chevy and the aftermarket. You can use this guide to compare the features and variables between the various small-block Chevy heads.

Manufacturer	Casting #	Port Volume Int/Exh (cc)	Valve Size (inches)	Combustion Chamber (cc)	Part Number
Chevrolet	3991492	161/61	2.02/1.60	64	3987376
Chevrolet	462624	161/62	2.02/1.60	76	464045
Chevrolet Bow Tie	14011034	184/55	2.02/1.60	64	10134392
Chevrolet	3932441	161/65	1.94/1.50	76	N/A
Chevrolet	3973487x	161/65	1.94/1.50	76	N/A
Chevrolet	3998993	159/64	1.94/1.50	76	N/A
Chevrolet	333882	160/60	1.94/1.50	76	N/A
World Prod. S/R	S/R	161/62	1.94/1.50	67 76	4351A or B 4350A or B
World Prod. S/R 305 Torquer	S/R	171/62	1.94/1.50	58	4205A or B
World Prod. S/R Torquer	S/R	171/62	2.02/1.60	67 76	4256A or B 4257A or B
Dart II Sportsman	N/A	200/N/A	2.02/1.60	64 72	1122A or B 1222A or B

Note: "A" designation after part number refers to a head with valves, springs, seals, retainers, locks and screw-in studs. The Sportsman heads also come with guideplates. "B" designation after part number refers to a bare head.

ALUMINUM CYLINDER HEADS 10

There's little that will get a true small-block enthusiast's heart pumping faster than a trick set of aluminum heads. Aluminum is the material of the future even for street engines and no engine is blessed with more alloy hi-po components than the small-block Chevy. As an example of this diversity, aluminum heads range in port volumes from 163cc production Corvette heads all the way up to fat-boy 220cc race heads. Just for street small-blocks there are at least five different companies producing 12 different aluminum heads. Add angle plugs, heat cross-overs, valve sizes and other variables and you have dozens of different ways to order a small-block aluminum cylinder head.

PORT VOLUME

Too many options can sometimes make choosing the right cylinder head a little difficult. The "bigger is best" theory rarely works for a number of reasons. While intake port volumes are given as a way to compare heads, the process is more complex than that. Port volume establishes a relative potential for port flow although there are usually airflow differences between two ports with the same volume.

Port Velocity

If maximum airflow was the only criteria, then those monster 220cc intake

I use the production L-98/H.O. 350 head as my starting point for many of the emissions-legal 350 and 383 cid EFI combinations. These heads sport a high velocity intake port, a small chamber for good compression and 1.94/1.50-inch valves stock.

ports would be the only heads anyone would use. But power is a nasty nest of compromises that require choosing the proper set of matching components. Port velocity is probably a term you have heard before. The inlet air/fuel mixture requires a specific velocity in order to deliver the maximum amount air/fuel mixture to the cylinder. As you can imagine, with a given test depression

(such as on a flow bench) a small intake port will generate a certain flow at a given velocity. Increase the cross-sectional size of the port and flow will usually (but not always) increase. Conversely, the larger port also decreases port velocity.

Condensing this down to the essentials, small intake ports tend to be very responsive and do a good job of filling

The Corvette L-98 head lends itself to excellent flow with some intelligent port work. Valve size is limited by the factory seat inserts to 2.00/1.56-inch valves but with the proper port work, these heads can generate up to 250 cfm of intake flow at .500-inch valve lift with an excellent E/I above 70 percent.

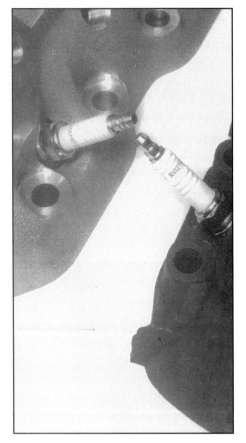

Many factory and aftermarket aluminum heads offer angle spark plugs. There is no real performance advantage to angled plugs and they often cause clearance problems with spark plug wires.

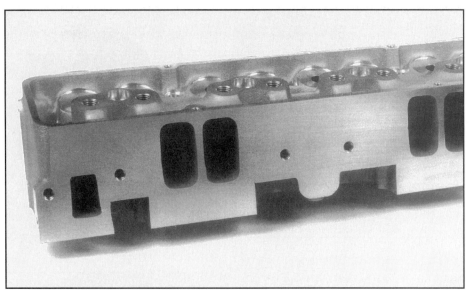

Here's a teaser for the ultra-killer group. Chevy's most popular race head is now the 18-degree aluminum head. The degree figure refers to the revised valve angle, which stands the valve more upright compared to the production 23-degree valve angle. This is the high-port version 18 degree head used in all GM Winston Cup race engines. The 18-degree valve angle change also requires a dedicated intake manifold.

the cylinder at mid-range and low engine speeds while large ports tend to be somewhat lazy at these same speeds. Larger ports are far better at generating efficient flow velocities at higher engine speeds where the smaller ports begin to choke. How big is a "small" intake port? I have enjoyed tremendous success with ported versions of Chevy's H.O. 350 aluminum cylinder heads on engines displacing as much as 383 inches. This head starts out with a small 163cc intake that porting enlarges to around 180 cc's. While this is a rough approximation, intake ports of around 180cc and smaller tend to improve intake port velocity in normal street driving rpm levels, making them desirable for an all-around street engine. Intake port volumes between 180 and 195cc's compose a mid-range group that offer an excellent compromise between good torque and peak rpm horsepower. The 200cc and larger group could be classified as big intake ports intended for high rpm use at the expense of low speed and mid-range power.

The mid-size heads with ports between 180 and 190 cc's are an interesting group. That's because heads such as Air Flow

Research's 190cc intake offer great promise for an everyday street engine. To underscore that point, the AFR 190 head currently enjoys a California Air Resources Board (CARB) exemption that makes it an emissions-legal substitute for virtually every small-block Chevy engine ever built. It will work especially well in 350 cid and larger applications but can also perform well in smaller-inch engines such as a 327 if high rpm power is what you seek. The Edelbrock Performer aluminum head with its 170cc intake port volume is another excellent street head that will deliver strong torque and horsepower.

Port Flow

As mentioned earlier, looking at port volumes alone is not enough. Port flow is actually more important. The ideal comparison would be to perform dyno comparisons of all the heads. But since cam timing plays such a big part in the power picture, it's difficult to choose one cam that would be "fair" to all the heads. A somewhat more equitable test might be a flow bench, but even this test cannot tell

WEIGHT BREAKS

One of the advantages of aluminum cylinder heads is a substantial weight savings. Before you even bolt them on, you can usually pull 50 pounds or more off the nose of your car. The following chart reveals the potential weight savings.

HEAD	WEIGHT (lbs)
Iron Production	44
Iron Bow Tie	46
Aluminum Corvette	19
Aluminum Brodix -8	22

COMPRESSION IMPRESSIONS

Compression is dependent on a number of variables. The following chart shows what happens to compression as chamber size changes. This example uses a 4.030-inch bore, 3.48-inch stroke (.030-over 350), a flat-top piston with -4cc valve reliefs and a .038-inch thick head gasket. We've also included various deck heights as another variable. The blueprinting chapter will outline the formula for computing compression ratio.

DECK HEIGHT (inches)	COMBUSTION CHAMBER VOLUME (cc's)				
	58	64	68	72	76
.025	10.68	9.96	9.54	9.16	8.81
.020	10.81	10.08	9.65	9.25	8.90
.015	10.95	10.20	9.76	9.35	8.99
.010	11.10	10.32	9.87	9.45	9.08
.005	11.25	10.45	9.98	9.56	9.17
.000	11.40	10.58	10.10	9.66	9.27

the entire story. However, we will use flow bench tests as a way to compare these different cylinder heads in the sidebar "Bench Test." As you can see, we have grouped the heads into the three different categories of Small, Medium and Large.

INTERPORT RELATIONSHIPS

One of the best ways to evaluate a cylinder head is what cylinder head specialists call the exhaust to intake (E/I) relationship. Exhaust port flow is expressed as a percentage of the intake port flow at a specific valve lift. For example, let's say we have an intake port that flows an excellent 250 cfm at .550-inch valve lift and an exhaust port with a weak 125 cfm at the same valve lift. Dividing the exhaust cfm of 125 by the intake flow of 250 produces a 50 percent E/I. In this case, the E/I indicates a poor

exhaust port. Remember to always compute the numbers at the same valve lift.

While a good intake port is always important, the exhaust port is critical to good horsepower. Most engine builders prefer a E/I relationship of between 75 and 85 percent as optimal. It's important to look at both intake and exhaust port flow in order to use the E/I percentage since a weak intake compared to a good exhaust port will also produce a "good" percentage that still won't make power.

While this may sound obscure for the average enthusiast, the E/I relationship does illustrate how each of the components in an engine relate to the final power the engine makes. A cylinder head with a great flowing intake port with good velocity will probably make great torque. But unless the head is teamed with an equally good exhaust port, it won't make great horsepower because at higher engine speeds, the exhaust port will not have time to dump all the residual exhaust gases from the chamber before the next cycle starts. This also

explains why a stock small-block Chevy cylinder head makes more horsepower when subjected to pocket porting. The exhaust port improves more than the intake, therefore improving the exhaust to intake relationship, especially at higher valve lifts.

Also keep in mind that cam timing plays a part in this relationship. A weak exhaust port can be helped with a longer duration exhaust lobe. A cylinder head with good E/I is usually better off with a single pattern cam.

COMPRESSION LESSON

Most hot rodders know that aluminum absorbs and transfers more heat faster than cast iron. This means aluminum heads tend to pull more heat out of the chamber during the combustion process. One tactic to compensate for this is to add static compression when converting to aluminum heads.

Since aluminum heads are often installed on engines originally equipped with iron heads, most cylinder head companies help by making the

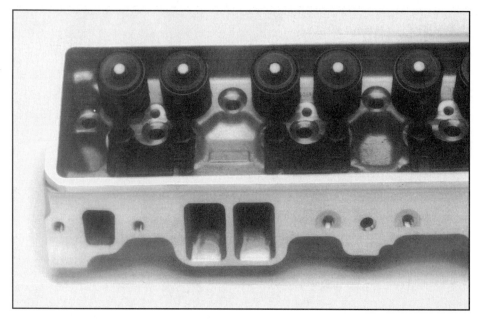

Edelbrock offers a centerbolt Performer cylinder head as an option for pre-LT1 '87-'93 Corvette and Camaro/Firebird owners looking for a better cylinder head. Of course, porting the existing aluminum heads is also an option, but for iron head F-car owners looking to step up, this is a great way to pump the power and lose a few pounds.

combustion chamber smaller. You can see from the cylinder head chart that many of the heads are offered with 68cc to 72cc chambers, down from the stock size of 76cc. Looking at the chart, you can see that reducing the combustion chamber by 4 cc's kicks the compression ratio by around .40 of a ratio. I typically will add a full point of compression for aluminum heads over an iron head from 9:1 to 10:1 on a carbureted engine. This allows maximum ignition timing on 92 octane

Edelbrock's Performer aluminum head has a 170cc intake port that offers excellent velocity and good port flow with a reasonable exhaust port. With the proper cam you can make over 400 lbs.-ft. of torque and 400 horsepower with these heads.

pump gas without undue detonation. If the engine is electronically fuel injected, the fuel distribution improves so the static compression can jump to as much as 9.8:1 with iron heads and 10.8:1 with aluminum heads! Note how improved fuel distribution to all cylinders allows the engine to run an overall higher compression ratio.

AFTERMARKET HEADS

If you recall the theory information presented in Chapter 1 (you did read it, didn't you?), you know that my emphasis on a wide powerband street engine calls for a cylinder head that will make torque even at lower engine speeds. This eliminates the 200cc and larger intake port heads, especially if the engine is a daily driven 350. Based on this concept, for a street engine of 350 to 383 inches, I prefer to use heads around 170 to 180cc intake ports. I use the aluminum Corvette head on many of my emissions-legal combinations because the head makes excellent power in a ported condition and because it is a bolt-on. Unfortunately, only a few of the aftermarket aluminum heads, such as the Edelbrock, AFR and TFS Corvette centerbolt heads will fit on a late-model Corvette without drastic modifications.

Edelbrock

According to early testing, a head that also works well in the under 180cc intake port range is the Edelbrock Performer RPM head. This casting combines a decent intake port flow with a good exhaust-to-intake port relationship. Since the intake port is only 170cc's, throttle response and low-speed torque are quite good. Plus, this head is available in a number of different configurations with angled or straight plugs, with or without a heat crossover, and it is available in an emissions-certified configuration in either the standard configuration or as a Corvette centerbolt style head. The Edelbrock Victor Jr. head is intended for

Brodix has the aftermarket aluminum cylinder head market covered with at least three cylinder heads ranging from the small Street Head, the 185cc Pro Street head and the venerable -8. The Smokey Yunick CNC-ported head shown here offers 2.02/1.60-inch valves ported by Weld Tech. This head is also completely emissions-legal (E.O. D-358). While expensive, the flow numbers are impressive.

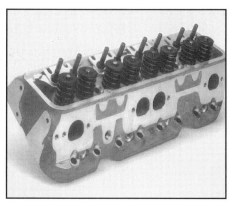

Air Flow Research offers two excellent street aluminum small-block heads. The 190 is intended for standard intake runner manifolds such as TPI and Edelbrock Performer style intakes. The 195 comes with a taller intake port for Victor, Jr. style intakes and some of the larger TPI style base manifolds. Either head is an excellent street choice. AFR also now offers the 190 version for centerbolt Corvette conversions and also a head to replace the new LT1 factory head.

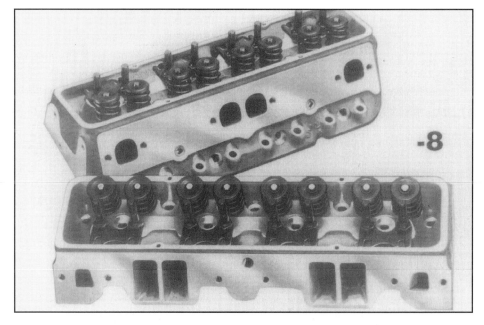

Brodix's -8 FSH Pro Street head is an excellent aluminum cylinder head. Fitted with 2.08/1.60-inch swirl polished valves, studs, springs, retainers and guideplates, the heads are ready to bolt on. With a 185cc intake port, it's a great dual-purpose cylinder street head.

serious, high rpm horsepower. The large, 212cc intake port would work well with high rpm 383, 406 or larger cubic inch small-blocks, especially in the 6500 rpm band. While this head can also be used on a 350, it will make the engine peaky and probably kill low and mid-range torque.

Brodix

Brodix recently added to the big mix of

aluminum small-block street heads with the Pro Street head. This head splits the difference between their smallish Brodix Street head and the -10 with a 185cc intake port and 2.08/1.60 valves and a 68cc chamber. This compares to the 194cc port volume of the -8 and the massive 210cc's of the -10 head. Beyond these street heads, Brodix offers the widest selection of aluminum race heads

for the small-block of any manufacturer. Brodix has literally dozens of heads to select from in the race head market. You can literally custom-build a set of heads from Brodix with variations in chamber size, port configuration and valve sizes. As with almost all of the companies, Brodix offers all their heads either bare or completely assembled, ready to bolt on.

AFR

Air Flow Research (AFR) offers two cylinder heads that fit the performance street market and the flexibility of these heads make them one of the better choices. The 190 head is intended for milder street engines and comes with a shorter intake port opening intended to match the Edelbrock Performer style intake manifolds. Both the 190 and 195 are also available in an emissions-legal configuration. The only difference in the 195 is its larger Victor, Jr. intake port opening.

The AFR heads are one of those interesting examples that appear to be borderline too large for an everyday street engine. However, experience with these heads on a number of my engines has proven them to be an extremely versatile

The LT1 (right) uses a completely different coolant pathway when compared to the typical small-block Chevy head. While it is possible to convert the LT1 to an earlier small-block the small flow advantage doesn't justify the cost of conversion.

This is a cross-sectional view of an Air Flow Research 190cc intake port. A taller short side radius (or port floor) tends to improve flow.

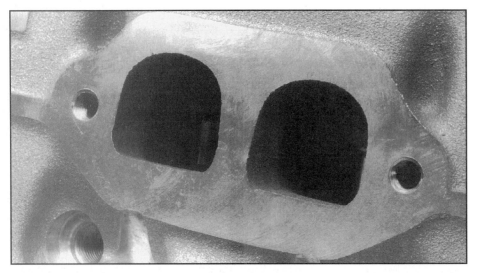

There are actually two versions of the Corvette aluminum heads. The first design '86 and '87 heads (casting number 128) used a standard exhaust port. Starting in '88, the newer head's (casting number 113) exhaust port shown here was raised .100 inch and given a "D" shape that improved exhaust port flow. Keep this in mind when shopping for used 'Vette heads.

HEAD SELECTION

As you can see, there are plenty of aluminum small-block heads to choose from. The choice really depends upon application. Everyday street engines will probably work best with smaller, high velocity intake port heads while engines destined for higher rpm use and more horsepower will benefit from larger 180 to 195cc intake port heads. But in any case, you can expect power to jump by just bolting on any of this group of aluminum cylinder heads.

THE 'VETTE HEAD SWAP

The late-model Corvette/H.O. 350 cylinder heads are quickly finding favor among the street set for their power, weight and cosmetic value. It's possible to run down to your local Chevy dealer and walk out with a pair of complete, ready-to-bolt-on aluminum heads for around $800 or even less from the mail-order dealerships like Scoggins-Dickey. However, that's not the whole story. These heads require careful thought and cost factoring before you lay down your long green.

Most importantly, these H.O. heads have a 58cc combustion chamber, which is substantially smaller than the standard 76cc iron head. On a typical .030-over 350 with flat top (-4cc) pistons, a .025-inch deck height and a .038-inch head gasket, the compression ratio with these

and powerful cylinder head. For ultimate torque on a 350, these heads may be a bit big, but for the 383 and 400 engines it is a great head that straddles the line between streetable torque and rpm horsepower.

Dart

Dart Machinery offers one of the older alloy heads with the 200cc Sportsman. This head teeters on the brink of being "too big" for the typical street small-block

although certainly many hot rodders have successfully used the head in street engines. For example, even on a 383 engine with a mild cam the 200cc intake port tends to slow the intake charge so that even this large small-block tends to lose mid-range torque because of lost intake velocity. On the other hand, this larger port tends to work better with a big cam for higher rpm power as long as soft bottom and mid-range torque is acceptable.

heads computes to 10.68:1, which is at least a 1/2 point too high to run with a carburetor on 92 octane pump gas, especially if you are using a short duration (less than 220 degrees at .050-inch tappet lift duration) camshaft. Even adding in Chevy's .051-inch head gasket, the ratio is still high at 10.34:1.

It's possible to open up the combustion chambers to decrease the compression. Increasing the chamber volume 6 cc's from 58 to 64cc's will drop the compression ratio back down to 9.96:1 with the .038-inch thick gasket. The other path if you are building an engine using new pistons is to use a half-dish piston (see Chapter 5) with an additional 6 to 10cc's that increase the total above-piston volume and create a more realistic compression ratio.

The H.O. heads also require a number of specific pieces to bolt it on. While the heads come complete and ready to bolt on, the heads are designed to use self-aligning, rail type rocker arms that center the rocker arm over the valve stem. The heads do come with pushrod guideplates, but they cannot be used with non-rail rockers since the guideplates are not hardened. This means you must use Chevy's Bow Tie aluminum head guideplates when using non-rail type rocker arms. Most aftermarket roller rocker arms will not clear the magnesium covers but Crane offers narrow-body roller rockers that will fit with some slight trimming to the valve covers. Finally, the exhaust port has been raised .100 inch and the spark plugs are angled, which may affect header-to-chassis clearance and proper spark plug boot fit around the headers.

The LT1

You may have noticed that the 1992 and later production LT1 aluminum cylinder head is conspicuously absent from this chapter on cylinder heads. While it will bolt up to any small-block Chevy engine, the LT1's new "reversed"

The new TFS Twisted Wedge head offers a new slant on the small-block head. With a 185cc intake port and a good exhaust port, this head is capable of good power. Even though the intake valve is moved from the stock 23 degree angle to 13 degrees, all stock valvetrain parts bolt on. TFS also offers similar Twisted Wedge designs for the Corvette centerbolt style and the new LT1.

cooling system makes it a unique head that doesn't work as a bolt-on. I have modified the head to adapt it for use on the normal small-block Chevy, but the flow bench numbers on the LT1 casting are only slightly better than the older L-98/H.O. 350 aluminum head. Plus, the LT1 employs an even smaller 53cc combustion chamber compared to the L-98 head's 58cc's. Because of the welding and machining required to convert the LT1, I don't recommend this head for use on older small-blocks. The L-98 Corvette or other high velocity heads would be a better choice and less expensive to install. ■

BENCH TEST

Comparing cylinder heads on the flow bench is a useful way to judge flow potential, especially when comparing heads of comparable port volume. A complete flow bench test from .100 to .600-inch valve lift is preferred but is beyond the space available here. As a result, we chose the .400-inch valve lift for both the intake and exhaust ports, with all tests performed at 28 inches of water test pressure on a SuperFlow flow bench. The E/I column is a comparison of the exhaust-to-intake relationship which expresses exhaust port flow as a percentage of intake port flow (see "Interport Relationships").

The problem with one data point for comparison is that all the other data points could be quite different between various heads even though they might flow the same at .400-inch lift. While this one data point was chosen as a mid-flow figure, you should do further research before deciding on one particular head. This data was obtained from out-of-the-box cylinder heads. Some heads offer hand blending of the seats while others don't.

Finally, these flow numbers were generated on different flow benches, although all were SuperFlow 600 units. There will be some variation in results due to differences in operator technique, data correction and other variables.

SMALL PORT	Port Vol. (cc's)	Intake (cfm)	Exhaust (cfm)	E-I (%)
Corvette/H.O. 350 *	163/58	186	152	82
Edelbrock Perf. RPM	170/60	219	151	69
Brodix Street	170/70	210	145	69
Chevy Bow Tie (Phase 6**)	175/59	192	126	66
MEDIUM PORT				
TFS	185/73	216	160	74
Brodix Pro Street	185/80	228	153	67
AFR 190	190/64	236	163	69
Brodix -8 ***	194/85	230	153	67
AFR 195	195/64	237	163	69
LARGE				
Dart II Sportsman**	200/NA	207	133	64
Brodix -10 ***	210/85	239	155	65

* The Corvette head is fitted with 1.94/1.50-inch valves while all the rest of the Small Port heads and most of the Medium Port are fitted with 2.02/1.60-inch valves.

** 2.05/1.60 valves

*** 2.08/1.6 valves

POPULAR ALUMINUM HEADS

The following is a short list of the more popular aluminum heads for the Mouse motor. This is hardly all of the heads that are available. We limited this chart to the more street-oriented heads and only list heads at 210cc or smaller intake ports. In addition to the specs listed here, there are additional options by various manufacturers including angled or straight plugs, heat cross-over, and extra-cost options of different chamber and valve sizes. For a more information, please contact the specific manufacturer.

Manufacturer	Casting Number	Port Volume Int/Exh	Valve Size (inches)	Combustion Chamber (cc)	Part Number
Chevrolet H.O. 350	10088113	163/58	1.94/1.50	58	10185086
Chevrolet Bow Tie Phase 6	14011049	175/59	2.05/1.60	55	10051167
AFR	N/A	190/64	2.02/1.60	68	N/A
				76	N/A
AFR	N/A	195/65	2.02/1.60	68	N/A
				76	N/A
Brodix Street	N/A	170/70	2.02/1.60	64	STPKG
Brodix Pro Street	N/A	185/80	2.05/1.60	69	-8FSH
Brodix -8	N/A	194/85	2.02/1.60	69	-8STD
Brodix -10	N/A	210/85	2.08/1.60	69	-10STD
Edelbrock Performer RPM	N/A	170/60	2.02/1.60	70	6063 6073
Edelbrock Victor Jr.	N/A	212/67	2.08/1.60	72	7720 7700
Dart II Sportsman	N/A	200/N/A	2.02/1.60	72	2212A/B* 2222A/B
TFS Twisted Wedge	N/A	185/73	2.02/1.60	63	TFS-31400001

* Dart II Sportsman offers the option of either straight or angled spark plugs.

11 CYLINDER HEAD MACHINING

There's a reason for spending so many chapters on cylinder heads. Since heads are an excellent source of power, it makes sense to cover as many advantages as possible for making inexpensive power. Choosing the right cylinder head is a good place to start. But unless you apply the right machining operations, it's very easy to turn a good cylinder head into an expensive paper weight. To put it simply, there is no substitution for quality machine work. If that means spending more money, so be it.

Much of what is covered in this chapter will relate more to reconditioning used heads than preparing new castings. Even the lowliest stock cast-iron head will respond to simple machine operations and modifications to improve port flow. Many of the operations detailed here also have as much to do with maintaining performance as they do with improving power. A quality rebuilt set of heads will last thousands of miles without deterioration. Cheap rebuilt heads are junk before the engine fires.

GETTING STARTED

When rebuilding cylinder heads, the first step is to ensure the head is worth rebuilding in the first place. Starting in the late '70s, Chevy began to lighten the iron cylinder heads to reduce weight. Lightweight casting such as the 882 and 624 heads are not particularly durable and

Checking for consistent valve seat height is one of the many head conditioning procedures you'll have to follow. This procedure should be done prior to the valve job, in order to establish consistent valve heights among all eight chambers.

often crack. The most common places for these cracks are around the exhaust valve seat and between the exhaust and intake seats. I always Magnaflux-inspect all used heads before machine work begins to eliminate the bad castings.

Pressure testing is another way to check for cracks or porosity. There are a number of different procedures, but I typically bolt the heads to a good block and pressurize the cooling system. If there is a leak, the pressure will drop. Another more expensive tool is a large fixture that seals and pressurizes the water jackets and then immerses the head in a tank of water to locate flaws. If you discover a crack in a production iron head, it's best just to look for another

Valve guides are one of the most important components in a cylinder head. There are several different styles from rigid cast iron guides (left) for use in cast iron heads to rigid bronze guides used in most aluminum heads (middle) to bronze guide inserts (right). Production Corvette heads use an iron guide.

Once new guides are installed, honing the guide establishes the proper guide-to-valve-stem clearance. Cast iron guides require slightly more clearance than bronze guides.

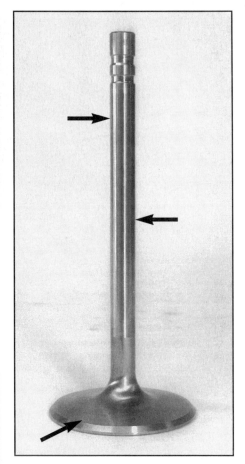

It is acceptable to reuse old valves if the stem wear is not more than .0005 inch. Stems usually wear at the top or near the bottom of the stem. Both exhaust and intake valves must have the proper margin thickness (arrow) of at least .050 to .080 inch. Grinding valves will reduce this margin. A too-thin margin will generate a hot spot and burn the valve. Exhaust valves require a thicker margin than intakes.

THE COST OF POWER

Everyone wants to get the most power for the money. One way to keep tabs on how much you invest is to honestly decide what power level you desire. Rebuilding a cast iron set of small-block heads correctly can be expensive. Between machining and parts, it's easy to invest $700 to $800 in a set of iron heads. Compare this to the cost of a new set of H.O. 350 or aftermarket aluminum heads such as the Edelbrock Performer RPM, Brodix or AFR heads, and you'll find these aluminum heads are only slightly more money. Since any of these aluminum heads will make more power than even a pocket ported set of production iron castings, it makes sense to spend a little more up front to make more power.

head. There are companies that can repair cast iron heads, but the procedure is expensive and there are no guarantees of success. Aluminum heads are more easily repaired with TIG welding with excellent results.

VALVE GUIDES

Even if your head rebuilding budget is limited, there are a couple of machining operations that are required. While some may prefer to invest in killer valves and trick springs and retainers, the key to a long lasting set of heads are the valve guides. Valve guides are the essential ingredient in any cylinder head because their job is to position the valve in

relation to the valve seat. If the guide is sloppy, the valve is allowed to move. While this may be only an additional .001 inch or so, that clearance is multiplied by the length of the valve at the seat. This allows the valve head to move around and strike the valve seat in a different place each time the valve returns to the seat. This quickly destroys the precise angles machined into the seat, ruining an otherwise good valve job. This in turn kills airflow followed immediately by a power loss.

A loose valve guide will also allow oil to leak past the valve guide seal. This oil contaminates the incoming air/fuel mixture resulting in lost power and

creating valve stem deposits that also contribute to lost air flow. All of this happens because of cheap, improperly installed or poorly clearanced valve guides.

Knurling—You may have heard of a procedure known as valve guide knurling. Knurling uses a machine tool to raise a portion of the inside diameter of the guide. This is an old technique often used in cheap rebuilds and should not be applied to a performance cylinder head rebuild.

Replacement Guides

There are a number of different types of valve guides from cast iron to bronze

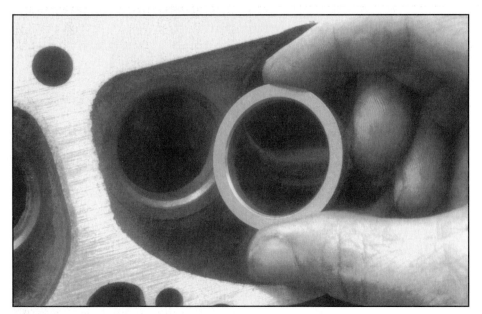

All small-block cast iron heads built after 1975 are fitted with induction-hardened seats. Earlier heads and these later heads that have been rebuilt more than once will require a hardened seat insert. This machining operation cuts into the original seat location allowing the machinist to drive in a hardened seat.

the exhaust since it runs hotter. Aluminum heads often use a solid, or rigid silicon/bronze guide although Chevy aluminum heads do come with iron guides.

New Valves—Ideally you should use new valves anytime the heads are rebuilt. If the valves are relatively new and the stems exhibit less than .0005 inch of taper or wear, then they can be reused. High quality stainless steel valves are typically much stronger and therefore last longer, paying off on the investment. A perfectly straight valve stem allows the machinist to run tighter clearances, which maintains the valve in the proper position to the valve seat, improving heat transfer to the guide especially for the exhaust valve. Tight clearances are better for heat transfer but can also create galling and a stuck valve causes more serious damage.

HARDENED SEATS

Beginning in the mid-70s, all new small-block Chevy heads were treated to induction-hardened exhaust valve seats as

wall guides. There are liners that fit inside the original guide and so-called rigid guides that are a complete new guide that is pressed into the head. Each machine shop has its preferences and no one style is right for all applications. Many quality machine shops prefer the bronze wall inserts since they leave a major portion of the original guide in the head for structural integrity. Of these, the interrupted spiral inserts are often the best choice. These guides offer a spiral groove cut into the bronze guide to position lubricating oil all the way through the guide. If the spiral formed a complete path from the top to the bottom of the guide, oil could find its way into the port. The interrupted spiral makes this path more difficult.

For cast iron cylinder heads, I prefer to use a cast iron guide. All small-block Chevy production iron heads come with iron guides and I like them because of their greater durability. New iron guides require drilling out the old guides and press fitting the new iron guide in place.

Clearances—Once the new guide is positioned, it is then honed to establish the valve-guide-to-stem clearance. This

clearance varies according to the type of guide used. For both cast iron and bronze guides on a street engine, I typically run .0012 inch for the intake and .0018 inch on the exhaust. More clearance is used on

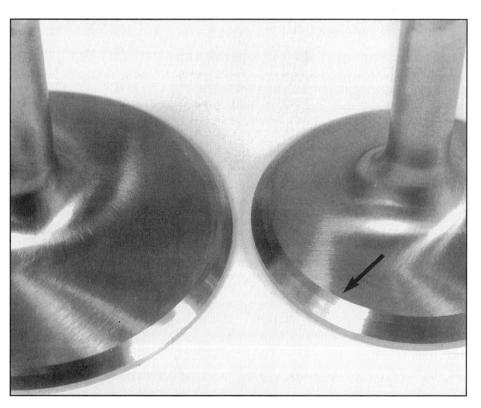

Machining a 30 degree back cut (arrow) on the inside of the 45 degree seat will improve low lift flow.

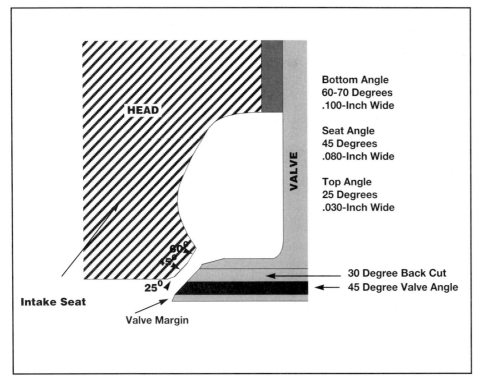

Bottom Angle
60-70 Degrees
.100-Inch Wide

Seat Angle
45 Degrees
.080-Inch Wide

Top Angle
25 Degrees
.030-Inch Wide

HEAD

VALVE

30 Degree Back Cut
45 Degree Valve Angle

Intake Seat

Valve Margin

This illustrates the relationship of a three angle valve job to the valve. Note the combination of angles tends to blend the airflow past the valve. This view also illustrates why a 30 degree back cut helps low lift flow. Placement of the seat contact point on the valve is also critical to flow. Note also the 25 degree top cut on the combustion chamber side of the valve.

I use a Serdi 100 to machine all valve jobs rather than using a grinder and stones. All 16 valve jobs are more consistent and the smoother finish creates a superior seal. The cutter is designed to machine all three seat angles simultaneously.

All aluminum heads should use some type of valve spring seat to prevent the spring from damaging the soft aluminum. There are two types of inserts, an O.D. cup that has a lip on the outside like this Brodix and an I.D. cup that locates the spring on the inside diameter.

a result of the use of unleaded gasoline. Lead in gasoline is an upper cylinder lubricant, so induction hardening was used to prevent exhaust valve seat erosion with use of unleaded gasoline. However, induction hardening penetrates only about .040 to .050 inch, which is good for

perhaps one valve job. Machining the new seats typically cuts away the hardened material, leaving the soft cast iron to quickly erode.

I generally remove the stock exhaust valve seats of any head we rebuild and replace the stock seat with a hardened

material similar to stainless steel. This material is especially tough and will withstand even the brutal demands of heavy towing loads over extended periods of time.

THREE ANGLE VALVE JOB

The three-angle valve job has become a performance standard that is almost universally accepted. What many enthusiasts don't know is that the factory usually only grinds the 45 degree seat angle and then merely machines a 70 degree throat angle intersecting directly with the 45 degree seat. This abrupt angle change from 70 to 45 degrees creates a shear condition that reduces airflow past the valve. By adding a 60 degree cut between the 45 degree seat and the 70 degree throat angle, the airflow is treated to a gentler radius that improves airflow.

POCKET PORTING

A great way to improve the port flow on a set of iron or aluminum heads is with pocket porting. Porting generally improves the air flow of any production head since production heads do not radius the port into the bottom of the valve job. Minor blending of the port reduces the sharp edges the air and fuel must pass over and improves the total airflow in the transition from the port to the valve.

Generally, the most beneficial place to port a small-block Chevy head is in the bowl area just beneath the valve seat. While many magazines have published stories on gasket matching intake ports to the intake manifold, this does little to improve airflow. As a general rule of thumb, the most difficult area to access in the head usually pays off with the most flow improvement.

While it's beyond the scope of this book to detail how to pocket port your heads, there are a couple of points worth noting. Merely hogging out the bowl area is not always ideal. I prefer to open the throat area just below the valve seat to around 85 to 90 percent of the valve diameter. This means that with 2.02/1.60-inch intake and exhaust valves, the bowl area directly underneath the valve should measure no more than 1.81 inch for the intake and 1.44 inch for the exhaust. Enlarging the bowl area more than this amount tends to reduce flow and also increases the chances of hitting the water jacket that surrounds the chamber.

Typically, the iron production small-block Chevy head responds to work on the exhaust side of the head more than the intake. If you were to measure before and after flow numbers for a given intake and exhaust port, you would see the exhaust generate a greater percentage increase compared to the intake. I have found a greater increase in top end horsepower versus low or mid-range torque when pocket porting iron heads. Keep in mind that pocket porting could further be enhanced with a camshaft swap to take advantage of the changes in airflow.

Pocket porting involves blending the port into the seat going no deeper than about one inch from the bottom of the valve seat. The usual procedure is to use a carbide cutting tool to rough-shape the port followed by abrasive hard rolls that smooth out the carbide cuts. Polishing this area is not necessary and results in no power gain.

While this may sound intimidating, it is actually easy as long as you work slowly and carefully. To do the job properly, you will need an air-operated die grinder, at least one carbide cutter, a handful of hard rolls, calipers to measure the bowl diameter, and safety glasses. You will also need some way to scribe a line in the pocket just below the valve job that establishes the throat diameter. This often falls just below the bottom cut of the three-angle valve job. The total tool investment will come to somewhere around $100.

A beginner can pocket port a pair of iron heads in less than 10 hours or I offer this service for virtually any cylinder head. Testing has shown that this effort is worth around 25 horsepower for a 9:1, 300 horsepower 350 small-block equipped with production iron heads. While some may feel this is more work than it's worth, if you are on a budget, there are few places where you can pick up this much power for so little money.

On the combustion chamber side of the seat, most shops also add a 30 degree top angle to create a more gradual radius to the chamber. This 30-45-60 series of angles then is referred to as the industry standard three angle valve job.

Most shops create these angles by grinding the seat with stones. This is the usual way to create the seat angles, but time consuming. A few years ago, the Serdi company created a machine tool that cuts all three angles simultaneously with a dedicated cutter instead of stones. The cutter not only cuts all three angles simultaneously along with the appropriate seat width but also creates a more consistent and smoother finish. I believe that the machined finish contributes to increased seat life while also eliminating high spots that sometimes occur.

Valve Heights—Of course, there's more than just angles to a quality valve job. Once any new seats have been installed, the first procedure before attempting the valve job is to establish consistent valve heights among all 8 chambers. This not only reduces chamber volume variation from cylinder to cylinder but also creates a more uniform valve stem height at the same time. This uniformity then requires fewer shims under the valve spring.

Sinking Valves—This is probably a good place to warn against sinking the valves. This is a more serious problem with used cast iron heads that have been subjected to numerous prior valve jobs. Each valve job removes a certain amount of seat material from the head. This sinks the valve deeper into the bowl area. This has a serious effect on airflow. Sinking the valve reduces the short turn radius of the port which hurts low lift flow. This is another reason to replace the seats if they appear to be deeper than stock. Installing new seats is a relatively simple procedure that any competent shop can perform.

Screw-in studs and guideplates are the classic modification to stock iron heads. The standard stud size is 3/8 inch while the portion that screws into the head is always 7/16 inch. To install the studs and guideplates, the machine shop must mill part of the rocker stud boss to accommodate the thickness of the guideplate. It may also be necessary to enlarge the pushrod hole in stock heads with higher lift cams to prevent binding.

Adding a positive valve seal requires the assembler to check retainer-to-seal clearance. Subtract the total distance from the top of the seal to the bottom of the retainer from the maximum valve lift. Minimum clearance is .050 inch.

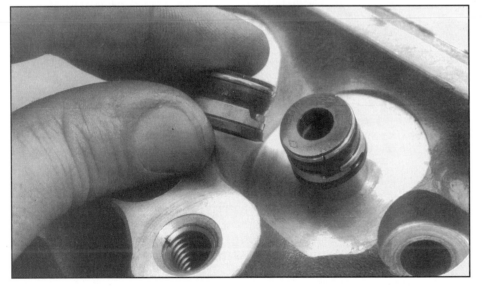

Positive Viton rubber valve stem seals from Victor are the only type of seal I use. These seals require the valve guide be machined down to .531-inch O.D. to install the seal.

Seat Width—Seat width is another important consideration. It's generally acknowledged that wider valve seats (especially for the exhaust valve) transfer more heat while the valve is closed, prolonging valve life. Conversely, a narrower seat width usually increases air flow. Drag race engines that enjoy frequent tear downs will sacrifice longevity for more power with thinner seat widths. For street engines however, thin valve seats merely pound out sooner and the flow goes down while also decreasing valve life. I prefer an .080-inch intake valve seat width with a .100-inch exhaust valve seat for street engines. To maintain seal, I machine the valve at a 44 degree 50 minute angle, 10 minutes short of 45 degrees. This slight interference angle maintains the valve seal.

Typically, moving the 45 degree seat out toward the edge of the intake valve also improves airflow. On a stock rebuild, the seat is usually in the middle of the valve. For better airflow and more power on street engines, I move the intake seat out to about .050 inch from the edge of the valve. Exhaust valves run hotter so I

Coil bind is another crucial clearance. Measure the height of the spring at coil bind as shown here. Then subtract that height from the installed height. Minimum coil bind clearance should be no less than .050 inch. Valve spring tension should also be checked and compared to the manufacturer's specification. It's always a good idea to use new valve springs with any rebuild, especially if high rpm operation is expected.

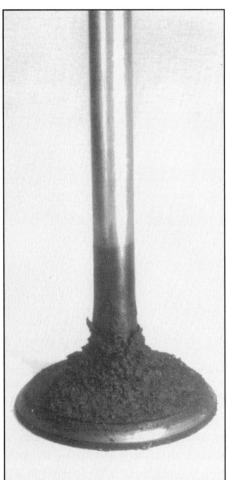

This is what an intake valve looks like when subjected to excessive guide wear and/or a failed seal. Oil is pulled past the guide and into the bowl where the heat cokes the oil on the back side of the valve. This kills airflow, power, mileage and throttle response. A leaking intake manifold gasket will also produce this type of buildup on the intake valves.

move the seat closer to the middle of the valve for durability. Ultimate street engines can move the intake valve seat out to within .020 inch of the edge of the valve, but the valve guide better be dead-on!

It's also important to mention that pocket porting any head does not pay off with improved flow unless you also pay attention to moving the intake valve seat out to at least .050 inch from the edge of

the valve. Flow bench testing has shown that an intake valve seat positioned in the middle of the valve can be improved as much as 10 cfm if merely moved to within .050 inch of the edge of the intake valve.

Every valve also has a margin which is the vertical height of the valve head just past the top of the 45 degree seat. Testing has determined that the thickness of this margin plays a supporting role in

improving airflow past the valve. It's preferable to start with a relatively thick margin, especially on exhaust valves. This enhances durability and improves flow slightly. Typical margin thicknesses run from .030 to .060 inch for the exhaust and slightly narrower for the intake valve.

Valve Sealing

I complete any valve job by checking the sealing quality of each valve job with a vacuum tester. This final step ensures that each valve is, in fact, sealed.

One of the final procedures for setting up a set of cylinder heads is cutting the top of the guides for positive oil seals. Production small-block heads use a

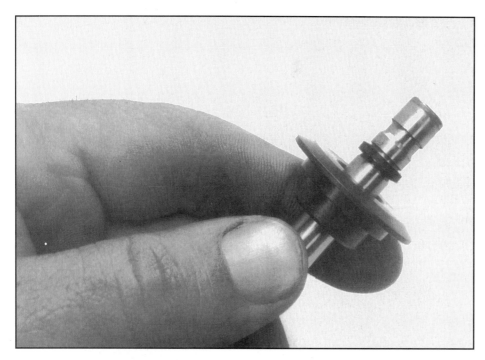

Many enthusiasts incorrectly install the stock o-ring seal. It must be installed after the retainer while the valve spring is compressed. Install the o-ring in the bottom groove of the valve stem and then install the keepers.

MILLING

Anytime either the block deck surface or cylinder head deck surface is machined, this changes the relationship of the intake manifold and cylinder head sealing surfaces. To restore this relationship and improve intake manifold gasket sealing, it's often necessary to mill both intake port sealing surfaces as well. If the heads and/or block is milled less than .020-inch, milling the intake is usually not necessary. All dimensions are in fractions of an inch.

| BLOCK/HEAD | INTAKE MANIFOLD | |
	Side	Bottom
.010	.012	.017
.015	.018	.025
.020	.025	.034
.025	.031	.043
.030	.037	.052

simple oil sealing combination of a thin oil shield called a shedder placed between the valve spring and the retainer combined with an o-ring placed over the valve stem after the retainer is installed. The combination of these two pieces prevents excessive amounts of oil from finding its way down the valve stem and into the combustion chamber.

PC Seals—While this system works relatively well, there are better ways of controlling oil past the valves. This is especially true with the intake valve since vacuum from that cylinder tends to pull oil past the seal when the intake valve is open. Many years ago the Perfect Circle company created what has become known as the PC seal. This is a hard plastic seal that fits tightly over a machined shoulder on the valve guide. The seal fits tightly over the guide and seals against the valve stem.

Flexible Seals—This seal is called a positive seal but is very inflexible. As the guide wears and the valve stem moves laterally, this movement elongates the hard plastic and allows oil to leak past the seal. Recently, Victor, Fel-Pro and other companies have created a rubber positive valve stem seal that works the same way as the original PC seal, but is much more flexible. The seal is made from Viton rubber, which is heat resistant and will remain compliant for long periods of time. Chevrolet uses this type of seal for the H.O. 350 aluminum head. I use the Victor R18E11F seal for most small-blocks. Brodix recommends using the original PC seals with their heads and bronze guide because the larger clearance is required because of the different expansion rates of the two materials.

The Victor positive seal requires machining the valve guide from its stock dimension down to .531 inch while others require a .500 inch guide. The small steel spring over the body of the seal retains the seal on the valve guide. A thin plastic shield is included with the seals that slips over the valve stem to make installing the valve stem seal easier. The outside diameter of this positive seal prevents the use of dual springs in a stock spring diameter cylinder head, but dual springs are rarely necessary for most street engines anyway.

As you can see, there's a world of difference between just throwing a valve job on a set of heads and attention to the small details that make a difference. The cost of doing the job properly understandably increases, but this is a wise investment if you are in search of a powerful and durable small-block. ■

12 CAMSHAFTS

If the cylinder heads are an engine's most important component, then the camshaft is a close second. The camshaft could easily be called the brain of any 4-stroke internal combustion engine since it controls the opening and closing of the valves. Your high school auto shop class probably oversimplified the opening and closing points, because the "bumpstick" is actually a complex piece of engineering and design. There are dozens of variables that come into play in camshaft design and application. As a result, there are hundreds of different cams that could be applied to the small-block Chevy. But based on the set of components used on any certain engine, there can be perhaps as few as two or three cams that would work best for that application.

Fortunately, making a slight error or not accounting for all the variables when choosing a camshaft is not seriously detrimental to engine power, especially for street engines. Erring on the conservative side of camshaft timing for street engines has many merits, making cam selection easier. We'll first take a quick jaunt through camshaft basics, and then investigate some of the more interesting facets of camshaft design that can make a big difference in creating power where you want it.

FLAT-TAPPET VS. ROLLER

The camshaft easily tops the list for the

The bumpstick is more than just a device used to open and close valves. It is actually a complex product of engineering and design. Proper camshaft selection can only be achieved when it is done with consideration of all other components—exhaust, heads, induction, etc.

most number of confusing terms to describe its functions. Each of these terms requires a solid understanding if you plan to decipher the number of specifications that are used to describe camshaft function. There are basically two different types of cam styles used in production and racing small-blocks.

Flat-Tappet

The flat-tappet cam is the most common, used in production small-blocks from 1955 through 1986. The flat-tappet cam is machined to utilize a tappet or lifter that has a somewhat flat face, giving the cam design its name. This tappet can either be a solid lifter or a hydraulic lifter. Both solid and hydraulic

It's easy to tell a flat-tappet cam (top) from a roller (bottom). Flat-tappet cams are machined from cast iron while the roller cam is usually machined from billet steel. Roller cams allow a faster opening rate which means they can offer more lift for the same degrees of duration, which can be a benefit in street engines. Rollers and flat-tappet cams are offered in both mechanical and hydraulic versions.

flat-tappet lifters have been used in production small-blocks. The solid lifter is just like it sounds, making a solid connection between the lifter and the pushrod. Because it's solid, there must be some design clearance, or lash, to allow for heat expansion. One result of this lash requirement is that the solid lifter cam lobe is designed with a clearance ramp built into the opening portion of the lobe to gently remove this lash from the valvetrain.

Hydraulic Lifter—The hydraulic lifter uses a small piston cushioned with oil inside the lifter that automatically compensates for heat expansion requiring no lash, no maintenance and no clearance ramp. However, hydraulic lifters can "pump up" at higher engine speeds, limiting engine speed. Therefore, high

rpm flat-tappet engines usually opt for solid lifter cams. Solid lifter cams, however, require regular maintenance to maintain this valve lash. Hydraulic lifter cams don't require this extra work, which is why hydraulic cams are the most often used for street engines.

Roller Cams

Roller cams are the second type of cam design, using a roller follower rather than a flat-tappet. As you might guess, the roller tappet rolls rather than slides across the cam face, generating less friction. This friction reduction results in a slight mileage improvement, which is why Chevy opted to utilize hydraulic roller cams beginning in 1987. This design requires a harder material for both the cam and the lifter, which is why all roller

cams are made of steel rather than cast iron in flat-tappet cams. Roller tappets can be either a solid or hydraulic design.

CAMSHAFT TERMINOLOGY

The most basic of camshaft specifications are lift and duration. Lift is fairly easy to determine, defined as the amount of rise given to the tappet measured from the base circle of the cam. This is called lobe lift, which is then multiplied by the rocker arm ratio, usually 1.5:1. Therefore, a lobe lift of .300 inch with a rocker arm of 1.5:1 would result in valve lift of .450 inch.

Duration

Duration is only slightly more complicated, defined as the amount of

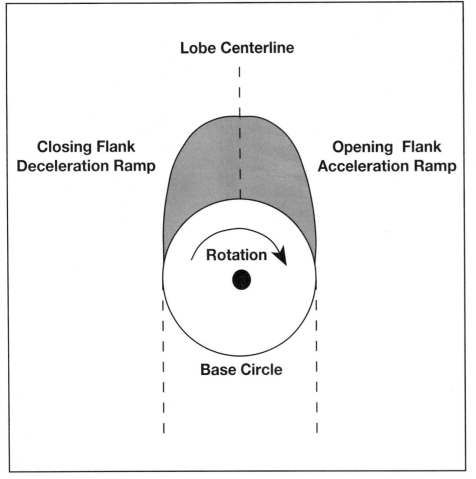

The first hurdle to understanding camshafts is to master the terminology. Lift is created by the height generated above the base circle of the cam. High lift cams require a smaller base circle so the lobe can clear the cam journals.

The lower roller cam has been Parkerized, a form of hardening, which gives it a black appearance. It is also a slightly different material. This cam also uses an iron distributor gear that is compatible with stock iron distributor gears. The upper steel roller cam requires a bronze distributor gear.

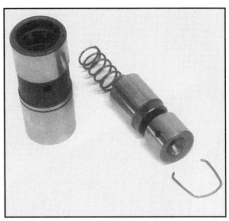

Hydraulic lifters are a very simple device utilizing a piston inside the lifter body. The area below the piston fills with oil and automatically removes any clearance in the valvetrain after the lifter is adjusted properly.

time, expressed in crankshaft degrees, that the valve is open. There are often two sets of numbers used to express duration. Advertised duration is the longer of the two duration numbers since it uses duration as measured from .004 or .006 inch of tappet lift. The second duration specification is identified as "duration at .050", using .050 inch of tappet lift as the opening and closing point for the duration specification. Since there are many different cam lift points for advertised duration figures, it's more accurate to use the .050 inch point when comparing camshafts, especially between different cam manufacturers. All hydraulic and solid flat-tappet cams can be compared at .050 inch, but you should not use the .050-inch figure to compare rollers to flat-tappet cams.

Lift

Lift is also a function of duration. For flat-tappet cams, there is a maximum amount of rate of lift than can be generated per degree of cam duration. This amount of lift-per-degree is limited by the diameter of the lifter. The larger the diameter of the lifter, the more lift-per-degree the lifter can tolerate. If the cam designer attempts to add more lift than the tappet can handle, the lifter will dig into the side of the cam and destroy

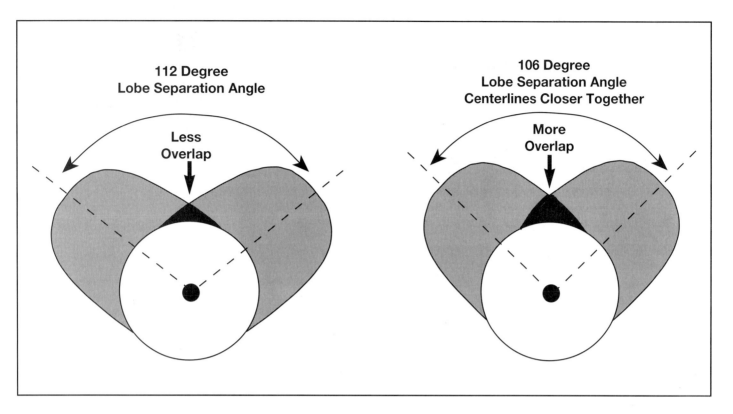

112 Degree
Lobe Separation Angle

Less Overlap

106 Degree
Lobe Separation Angle
Centerlines Closer Together

More Overlap

Imagine looking at a camshaft straight down the shaft and this is how the lobes would look. Lobe separation angle is the angle between the intake and exhaust lobe centerlines. As you can see, if the lobes are pulled closer together, overlap increases and the lobe separation angle becomes smaller. Conversely, if the lobes are machined further apart, the lobe separation angle becomes larger and overlap decreases.

Roller lifters used to be expensive, race-only items. Now even Chevrolet uses hydraulic roller lifters in production small-blocks. The factory hydraulic roller lifter (left) is heavy and can only be used in factory post-'86 blocks because of its height. Comp Cams offers a replacement roller lifter for these applications as well. For flat-tappet cam cylinder blocks, Comp Cams (shown), Speed-Pro and Crane offer a hydraulic roller lifter (middle) that drops right in. The most popular rollers are still the lightweight mechanical such as this Comp Cams roller lifter (right).

both the lifter and the cam.

A roller lifter has a much higher lift-per-degree limitation because of the roller cam follower. That is why a roller cam will generate much more lift for the same amount of duration. This is an inherent advantage that a roller cam has over a flat-tappet cam. For example, a Comp Cams 280HR hydraulic roller offers .525-inch lift versus .480-inch lift for a 280H flat-tappet hydraulic cam while reducing the duration at .050 inch by 6 degrees! In the world of no free lunches, however, roller cams and their required valvetrain pieces are more expensive than flat-tappet cams.

A production small-block Chevy flat-tappet hydraulic cam is generally around 200 degrees of duration at .050-inch tappet lift for both the intake and exhaust. Each lobe therefore has an opening and closing point expressed in relation to Top Dead Center (TDC) of cylinder number one. Intake opening (IO) is expressed as the number of degrees Before Top Dead Center (BTDC) while Intake Closing (IC) is the number of degrees After Bottom Dead Center (ABDC). Exhaust opening (EO) is given as the number of degrees Before Bottom Dead Center (BBDC) while exhaust closing (EC) is stated in the number of degrees After Top Dead Center (ATDC).

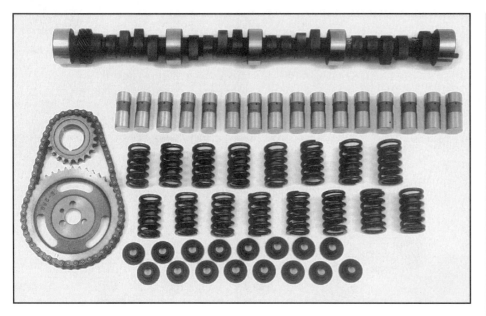

The best plan when buying an aftermarket hydraulic roller cam for pre-'87 small-blocks is to purchase a complete kit. This photo is of a flat-tappet Comp Cams K-kit but the kit concept is the same for hydraulic rollers. The K-kit offers everything you need including lifters, springs, retainers, locks, seals, timing chain set and assembly lube. The hydraulic roller cams can also be ordered from Comp Cams with a special iron gear and rear journal, allowing the use of a stock iron distributor gear.

Checking Duration

Let's say you discover an unknown cam sitting on the shelf. If you have the opening and closing points of the cam, here's how to determine duration. Add the opening and closing points of the intake together and add 180. For example, IO = 28 BTDC and IC = 60 ABDC so 28 + 60 + 180 = 268 degrees. The same is true for the exhaust. This works for both the intake and exhaust. You can use this trick for determining camshaft duration for cams that you may not have the specs for. The only way to determine these opening and closing numbers is to place the cam in the engine and measure the opening and closing points. The procedure for degreeing a cam is explained in Chapter 17 on *Blueprinting*.

Intake Centerline

This next spec should be easy to remember. There is a centerline at the exact peak of the intake lobe. Like the opening and closing points, intake centerline is expressed in relation to TDC. A common point for the intake centerline on most small-blocks is between 106 to 110 degrees ATDC. Cam manufacturers like Comp Cams often use intake centerline as a common reference position to indicate the position of the cam in relation to TDC. When you degree a camshaft, moving the camshaft relative to TDC changes all the opening and closing points as well as intake centerline.

You've probably heard the term "advancing" or "retarding" the cam. Intake centerline is an easy position from which to determine position of the cam. Let's say that you've installed a Comp Cams 268 at its "straight up" intake centerline of 106 degrees ATDC. Later, you decide to advance the cam two degrees, which tends to improve bottom end torque. Advancing the cam means moving all opening and closing points two degrees sooner. This also changes the intake centerline from 106 to 104 degrees ATDC. Retarding the cam two degrees from its 106 centerline would place the intake centerline at 108 degrees, opening and closing the valves later. This generally improves top end power at the expense of some low end torque.

Lobe Separation Angle

Unlike intake centerline which can be changed when installing the cam, lobe separation angle is ground into the camshaft when it is machined and cannot be changed without grinding a new cam. Lobe separation angle is the angle between the intake centerline and the exhaust centerline expressed in crankshaft degrees. This angle is one way to express the amount of overlap between the intake and exhaust valves. As you know, the intake valve opens slightly before the exhaust valve closes. This overlap is affected both by the length of

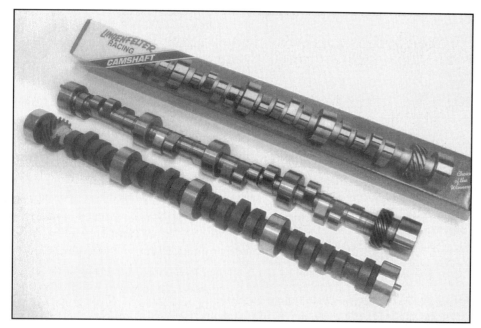

I have designed and developed five different camshafts for the Tuned Port Injected (TPI) engines. Two are flat-tappet hydraulics while the other three are hydraulic roller cams. These cams are not compatible with stock GM PROM's, so an aftermarket chip is recommended or a stand-alone electronic fuel injection control, such as ACCEL's EFM system, can be used. The hydraulic roller cams also feature a cast iron gear pressed on to the steel shaft to allow the use of a stock distributor gear.

duration of both lobes as well as each lobe's centerline. If you refer to the lobe separation angle illustration, this will become more clear. As you "tighten" or move the lobes closer together, the lobe separation angle becomes smaller. This moves the angle, for example, from 114 degrees to 110 degrees. Conversely, it can be moved further apart, such as moving the angle from 106 degrees to 110 degrees. The range of lobe separation angle is usually from 106 to as much 116 degrees but most performance small-block cams are ground between 106 and 114 degrees.

Changing Valve Overlap

But there is more to lobe separation angle than just valve overlap. The reason for changing valve overlap is to adjust the power curve within the power range dictated by the cam's duration. For example, factory camshafts limit valve overlap by widening the lobe separation angles to as much as 114 to 116 degrees. This improves idle quality, but it hurts low and mid-range power. As an

example, let's take a 210 degree single pattern cam with a 114 degree lobe separation angle and tighten the lobe separation angle by moving intake and exhaust lobes tighter together to 110 degrees.

Idle quality will immediately drop 2 to 4 inches of manifold vacuum from 18 inches to perhaps 14 inches or less. This will give the engine a slight lope at idle. More importantly, the torque curve from 1500 rpm up through 4000 rpm will increase because we're opening the intake valve sooner. The downside is perhaps a slight decrease in power at or above the peak horsepower point. For a street-driven engine this might be a good trade-off since we're still working with a short duration camshaft with excellent throttle response. Even tightening the lobe separation angle from 114 to 112 can make a significant difference in torque.

Lobe Separation & Longer Duration—Where lobe separation angle becomes confusing is when you deal with longer duration camshafts. A 110-degree lobe separation angle may not generate great amounts of valve overlap with a short duration 210 degree at .050 cam. But choose a 250 degree at .050 camshaft and a 110 lobe separation angle creates much more valve overlap. In long duration cams, it's possible that a wider lobe separation angle will create a wider

Most billet steel solid roller cams require a silicon bronze distributor gear. If the stock iron distributor gear is used with a steel roller cam gear, the iron gear will quickly wear, ruining the cam gear as well. The softer bronze gear prevents damage to the cam. The bronze gear will not live indefinitely, but its life is increased if used with a standard pressure and volume oil pump since the distributor gear also drives the oil pump. High pressure/volume pumps increase load on the distributor gear.

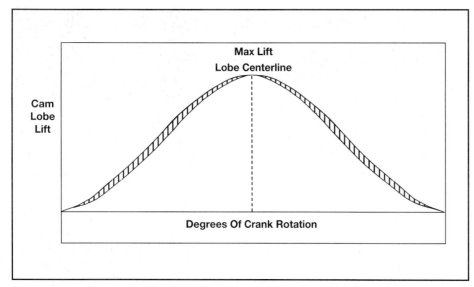

Two cams can have the same lift and duration but vastly different profiles. Note the shaded area illustrates how one profile can be much more aggressive on the flanks than another even with the same seat timing and lobe lift.

power band with more top end power. Of course, the next engine might respond more favorably to a tighter lobe separation angle with this same duration. Ultimately, experimentation is the only way to determine the best camshaft for best power.

There are several variables that come into play with lobe separation angle and they can be somewhat confusing. You can move the intake lobe relative to the exhaust lobe, which will change both the intake centerline and the lobe separation angle. Or you can move just the exhaust lobe, which won't change the intake centerline but will change the lobe separation angle. Or you can move both the intake and exhaust lobes. As you can see, there are plenty of variables to experiment with just within a single lobe profile.

Patterns & Symmetry

In the early days of hot rodding, virtually all cam lobes were symmetrical. The opening and closing flanks of the cam lobe were identical and the intake lobe profile was also used on the exhaust. While this works in many applications, the cam companies have long known that the lobe does not have to be symmetrical

in shape. In fact, most modern day race cam designs and many street cam designs are asymmetrical, often using a less aggressive closing ramp to prevent the valve from bouncing off the seat at high rpm. This is called an *asymmetrical lobe profile*.

A different exhaust lobe profile and duration can have dramatic effects on engine power. As indicated in the cylinder head chapters, even aftermarket performance cylinder heads often suffer from a weak exhaust port. This is especially true with production cylinder heads. By adding duration to the exhaust lobe, the cam designer can help that weak exhaust port by giving the exhaust port more time (in degrees of exhaust lobe duration) to vent exhaust gases from the cylinder. A camshaft with a different exhaust duration compared to the intake is referred to as a dual-pattern camshaft.

FOCUS ON DURATION

If you were to focus on only one cam spec, it would have to be duration. Intake duration positions the rpm range where the engine will perform at its best. Short duration camshafts make good torque since they open and close the intake valve over a short period of time when engine

speed is slow, when there is plenty of time to fill the cylinder. Unfortunately, this same short duration does not allow the cylinder time to fill at higher engine speeds where there is less time to fill the cylinder.

Conversely, a longer duration intake lobe allows more time for cylinder filling at higher rpm, creating more power. The down side is that this duration is too long at lower engine speeds. Usually, this longer event duration closes the intake valve much later allowing the piston to actually push a portion of the intake charge past the intake valve at low engine speeds. This, combined with exhaust dilution of the incoming fresh charge is what gives a long duration camshaft-equipped engine that choppy idle quality.

Intake Closing—Among the four valve events, the intake closing (IC) point is by far the most important. This point determines how far up the cylinder the piston travels before the intake valve closes. Close the intake valve early and the engine makes good torque but cannot fill the cylinder adequately at higher rpm. Close the intake later and low to mid-range torque will suffer since that later closing valve will allow air and fuel to be pushed back up into the intake tract.

Exhaust Opening—Exhaust opening (EO) is the second most important point. When combined with the IC point, EO establishes the lobe separation angle. Open the exhaust valve later and low-speed torque improves while opening the exhaust valve sooner allows the cylinder to dump spent gases earlier, helping high speed operation when combined with longer duration. The remaining IO and EC points have less of an effect on engine power than do IC and EO.

CAM SELECTION

When choosing a cam, the engine's intended usage and its displacement are critical. A mild, everyday street-driven 350 will want a shorter duration cam of around 210 to 215 degrees duration at

Roller cams are becoming increasingly popular with the street set. Don't be afraid of using a mechanical roller cam for a street engine since these latest cams tend to maintain their lash, decreasing the need for constant adjustment.

Always coat a new cam with the manufacturer's recommended break-in lube before using it the first time. The most critical time for a new cam is the first 20 minutes of operation. Start the engine and immediately bring the engine speed up to at least 1500 to 2000 rpm for the first 15 to 20 minutes.

.050 to enhance low and mid-range torque where the engine will be operated most of the time. However, there is a difference between a mild street cam for a 283 and one for a 406-inch small-block. In general terms, the larger the engine, the more duration it can withstand and still make good power. For example, if all we did was switch from that 350 to a 400, we could pump the cam to around 220 to 224 degrees of duration at .050.

Stroke

Stroke also has a significant affect on cam selection. Increased stroke, such as a 383's 3.75-inch stroke versus a 350's 3.48-inch stroke, allows the engine builder to increase duration slightly since that longer stroke tends to decrease sensitivity to cam duration. For example, a 222 degree at .050 cam even with a tight 110 degree lobe separation angle will still work well in a carbureted 383 for everyday street driving, especially if combined with a Performer RPM dual plane intake. A 406 small-block could handle even a couple of degrees more intake duration and still be streetable. Of course, these engines do lope slightly at idle. A wider 112 to 114 lobe separation angle would smooth the idle but hurt mid-range torque somewhat.

Cam Secrets

Choosing a cam can be confusing. There are well over a dozen different cam companies all claiming their cams make more power than the competition's. For a street-driven small-block Chevy, each of these companies offers more than a handful of different cams. Once you've narrowed the selection process down to duration within 4 to 6 degrees, how do you pick the right one?

When I have two cams of the same duration, I will choose the cam with the most lift. For example, Cam 1 is the mid-70s L-82 factory Corvette hydraulic cam with 222 degrees of duration at .050 and .450-inch lift. A second cam is a Comp Cams 275 Dual Energy hydraulic with 219 degrees of duration and .460 lift. A quick comparison of each cam's duration and lift reveal that the slightly shorter Comp Cams 219 cam has a more aggressive lift curve. Most aftermarket camshafts are new enough that there are insignificant differences between the lift curves. However, older designs like this Chevy cam used more duration to create less lift. In this comparison, the shorter duration 219 degree cam will have a crisper throttle response and more torque compared to the Chevy cam. In fact, Chevy's advertised duration is listed at 312 degrees where Comp Cam's advertised duration is 37 degrees shorter at 275! Shorter valve timing with more lift for street cams is the way to go!

CAM SPECS

The tremendous number of variables in engine combinations prevents any type of overall camshaft recommendation chart. However, I have tested dozens of Chevrolet Tuned Port Injection (TPI) engines and have come up with five camshafts that work very well with either modified TPI intakes or with the SuperRam intake manifold. While all the cams are designed to be used with EFI intakes, they can also be used successfully in carbureted street engines.

Hydraulic Flat-Tappet Cams

Part Number	Duration @ .050 I/E	Valve Lift (inches) I/E	Lobe Separation (degrees)
74212	214/224	.443/.465	112
74213	224/234	.465/.488	112

Hydraulic Roller Cams

Part Number	Duration @ .050 I/E	Valve Lift (inches) I/E	Lobe Separation (degrees)
74211	211/219	.503/.525	112
74216	216/219	.462/.470	112
74219	219/219	.503/.525	112

• The 211 or 212 cams are designed for use with stock or High Flow runners, offering strong mid-range torque and a smooth idle

• The 213 or 219 cams are intended for 383/406 cid engines in conjunction with the SuperRam intake. These cams have a choppier idle.

• The 216 hydraulic roller offers less lift that does not require valve guide machining to the heads. High Flow TPI runners are suggested and this cam will produce a smooth idle with excellent torque.

• Note the cams designed for the larger 383 and 406 cid engines still don't utilize duration above 224 degrees for the flat-tappet cam and 219 degrees for the roller.

A WORD ABOUT REVERSION

If you've ever seen black, sooty carbon coating the walls of an intake manifold or inside the carburetor, then you're looking at reversion. This happens most often to street engines that spend most of their time at speeds below 3000 rpm. The major cause of reversion is long duration camshafts combined with too much valve overlap.

In the four stroke cycle, the intake valve opens before the exhaust valve closes. The time when both valves are open is called overlap. A long duration camshaft opens the intake valve sooner into the tail end of the exhaust cycle. At low engine speeds, below where this duration is intended to work, the intake valve opens into significant exhaust pressure in the combustion chamber, much higher than in the intake manifold. This higher pressure exhaust gas forces its way past the intake valve and into the intake manifold. At some point, since the exhaust valve is also open, the pressure in the intake, chamber and exhaust pipe are equal and the exhaust gas upstream travel in the intake stops and then begins to retreat back toward the chamber. In the worst case, the exhaust gas does not travel back down the intake until the piston begins its downward, intake stroke as the exhaust valve closes. This traps some of the exhaust gas in the chamber, diluting the fresh intake charge. This exhaust gas dilution contributes to the rough idle characteristics of a long duration engine.

The farther this soot travels up the intake manifold, the worse the reversion problem. This is accompanied by terrible gas mileage and a sluggish engine. Other contributing factors to reversion can be a larger intake manifold like a tunnel ram and overly large intake ports.

The best cure for reversion is a shorter duration camshaft that will open the intake valve later, after the exhaust gas has had a chance to escape out of the chamber. This is especially true for street-driven engines that spend most of their time at engine speeds below 3000 rpm.

CAM-CLUSIONS

There is much more to cam design and application than we could possibly cover in this short chapter. Thankfully, most of this science is handled for you by the cam manufacturer. Once you've gained some camshaft experience and how cams interact with the rest of the engine, you'll find that camshafts are not as intimidating as you once thought. One of the best parts of all this is that most camshafts are not that expensive, so experimenting can be educational without killing your wallet. ■

HOW TO READ A TIMING CARD

Every aftermarket camshaft includes a spec sheet called a timing card. While each company prints their cards differently, most of them detail the same information. The timing card lists specific opening and closing points for both lobes as well as other information that is usually not listed in the catalog. The timing card information is useful for both reference and to use when degreeing in the camshaft. We will use a Comp Cams timing card for this example. The numbers below correspond with those on the chart.

1. Camshaft part number and application
2. Grind Number is often different from the part number. This is usually the number stamped on the end of the camshaft.
3. Valve Adjustment will specify a lash clearance if the camshaft is a mechanical cam. In this case, the cam is a hydraulic which requires lifter preload.
4. Gross Valve Lift is determined by multiplying the lobe lift times the stock rocker ratio of 1.5:1. In this case, Comp Cams has rounded the lift off to .460 inch (.3026 x 1.5 = .454)
5. Duration At .006 Tappet Lift is the duration measured between .006 inch of tappet lift on the opening and closing side. Compare this to Duration at .050.
6. Valve Timing is given on this sheet at .006 inch tappet lift for the intake and exhaust opening and closing points. Other cam companies such as Crane list the opening closing points at .050 inch. Note also that these opening and closing points are for the cam installed with a 106 degree intake centerline. Moving the centerline will move the opening and closing points as well.
7. Duration at .050 is the duration of each cam lobe at the .050-inch tappet lift checking height.
8. Lobe Lift is the actual lift as measured at the lobe.
9. Lobe Separation is the angle in cam degrees between the intake and exhaust lobe. This is one way of referencing valve overlap.
10. In this case, Comp Cams also makes a recommendation for matching Comp Cams valve springs that complement this camshaft.

1—PART # 12-210-2
ENGINE: CHEV. SML BLK 265-400

2—GRIND NUMBER: CS 268H-10 HIGH ENERGY

	INTAKE	EXHAUST
3—VALVE ADJ.:	HYD.	HYD.
4—GROSS VALVE LIFT:	.460	.460
5—DURATION @		
.006 Tappet Lift:	268°	268°

6—VALVE TIMING:

@ .006 Tappet Lift	Open	Close
Intake:	28° BTDC	60° ABDC
Exhaust:	68° BBDC	20° ATDC

These specifications are for a cam installed at 106° intake center line

	INTAKE	EXHAUST
7—DURATION @ .050:	218°	218°
8—LOBE LIFT:	.3026	.3026
9—LOBE SEPARATION:	110°	

10—Recommended revalve spring no. 981 valve spring specs. furnished with springs.

13 VALVETRAIN

Valves and the attending valvetrain components are essential items in the horsepower equation. Each and every component has a specific job. As with most other engine parts, you can choose stock pieces and the engine will run fine. It might even make decent power. But once you begin the search for more power, this leads to specialized pieces that contribute to pushing the peak of that power curve. We'll look at a number of these aftermarket components. None of these pieces individually will magically produce great quantities of torque or horsepower. But combine many of the improvements these pieces promise and significant power gains can be realized.

Many of the advances in valvetrain technology over the years are based on increasing durability as rpm levels reach ever higher. In most race engines, rpm is limited not by the bottom end, but by the valvetrain, specifically the weight of the valves and the strength of the valve springs. Valve float, which is discussed later on in this chatper, is a dangerous condition that destroys valve spring pressure and ultimately can contribute to a dropped or broken valve, which is never fun no matter at what rpm it occurs.

These valvetrain advancements have filtered down to the street market from what were once exotic, race-only pieces to affordable street pieces. The combination of a set of properly

As you push the rpm envelope higher, you must increase the quality and durability of the valvetrain. Rpm is limited in large part by the capabilities of your valvetrain components.

assembled heavy-duty valve springs, roller rockers, lightweight retainers and proper length pushrods can produce a street small-block that is both powerful and extremely reliable.

VALVES

The small-block Chevy valve market is a sea awash with choices. Where valves used to be limited to TRW or Sealed Power stock replacement pieces, there are now dozens of companies offering a

bewildering selection of valves. It's far beyond the scope of this short chapter to detail all the differences in valves from all the different companies. Instead, we'll look at some general differences in valve materials and shapes to help you make sense out of all the different offerings.

Stock small-block valves are typically produced out of an alloy carbon steel with only passing regard for optimizing flow. These valves and their aftermarket replacement brethren are fine for an

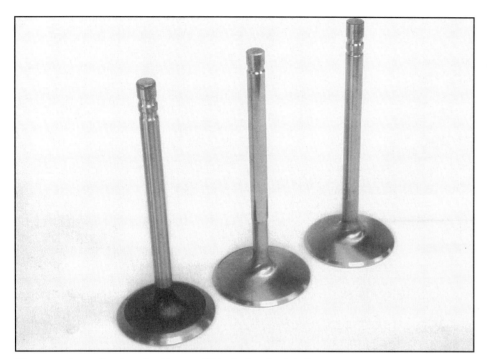

There are tons of valves for the small-block. On the left is a stock replacement 1.94-inch intake. The middle valve is a necked-down style 2.020-inch intake valve. On the right is a Lingenfelter stainless steel 2.00-inch intake valve that is used on Corvette heads. This is the largest valve that can be used with the stock 'Vette seat inserts.

Valve margin is defined as the thickness of the valve head itself. Thin margins reduce valve weight, but a too-thin margin can cause valve failure. I like to see the intake margin no less than .040 inch and no more than .070 inch. Exhaust valves are usually thicker at between .060 and .090 inch. A thicker margin tends to improve low- and mid-lift flow.

everyday street driver looking for inexpensive pieces that will last for 70,000 miles or more. While inexpensive, stock replacement valves tend to limit performance from an airflow standpoint. When subjected to higher rpm and higher exhaust temperatures these valves are basically out of their element.

Valve Sizes & Choices

There have been a number of different valve sizes applied to the small-block Chevy from the tiny 1.72/1.50 inch, up through 2.02/1.60 inch with the 1.94/1.50 inch intake and exhaust the most common. The largest production exhaust valve is the 1.60-inch piece although some aftermarket heads can squeeze in slightly larger 2.05/1.65-inch intake and exhaust valves. It's not always true that the cylinder head with the largest valves is the best. Usually, excellent street performance can be gained through using the popular 2.02/1.60-inch valve combination.

When you dive into the stainless steel valve market, you also discover a tremendous variation in stainless

materials. Typically, as the cost increases the material becomes stronger and more durable. For example, Manley offers five different material choices of stainless steel valves and within each of these types they offer numerous stem and valve head diameters, necked-down stems, tuliped valves and different valve lengths! In other words, literally hundreds of different variations.

I have also done my own stainless steel valve research for the 2.00/1.56-inch valves used in the aluminum Corvette heads. These are the largest valves that can be squeezed into the production seats. Larger valves could be used but would require expensive valve seat machining. These valves come from Argentina and my independent testing has shown that not only is this material as strong as similar stainless alloys used by other valve companies, but that it's also competitively priced. For other performance street engine applications, I typically use a Manley or Ferrea stainless steel necked-down valve typically in a 2.02/1.60-inch combination.

Among performance valves, the

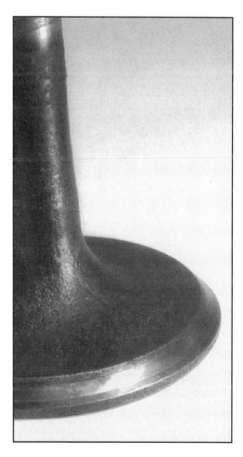

The indentation behind the 45 degree seat on a production valve can affect low-lift flow on any small-block head. The best way to eliminate this is to perform a 30 degree back cut in this area. This back cut removes the indentation and adds a radius to improve flow.

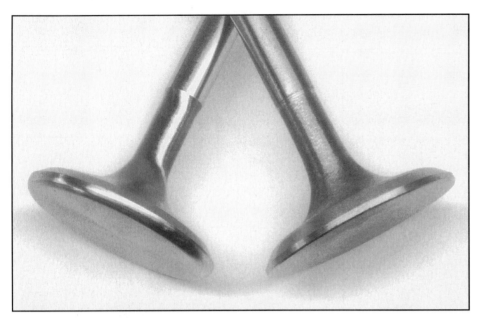

Some shops will actually radius the top of the exhaust valve in an effort to improve flow on the exhaust side. This radiused exhaust valve comes stock in the TFS head.

acceleration rate camshaft profiles. Any combination of these factors can quickly limit even new but stock valve springs to perhaps as little as 5500 rpm. Chevy realized this with the new LT1 engine and improved its valve spring, giving it a higher rate of 373 pounds per inch versus older spring's 267 pound per inch figure, even though the seat pressures are roughly the same. Chevy also lightened the retainers to help rpm potential.

Single vs. Dual Springs

It's important to know the difference between single springs with a damper and true dual springs. Typically, the small inside diameter of a production small-block spring precludes the use of a second, inner spring. However, even stock springs use a flat wire spring, called a damper, inside the main spring. This flat wire is used to dampen the main valve spring oscillations at a particular spring frequency to ensure long spring life. The addition of this flat wire damper is not considered to make the spring a dual spring.

A true dual spring actually has a smaller, round wire spring inserted inside

necked-down stem valve has become very popular. This style valve takes a standard small-block 11/32-inch (.3415-inch) valve stem and undercuts the stem below the guide in the valve bowl area. This reduced stem thickness is claimed to increase flow, especially at low valve lifts. There are some slight improvements in airflow in certain applications. However, claims of huge 40 percent flow increases with merely the addition of an undercut stem valve are probably exaggerated. Similar tests using these valves on my SuperFlow 600 flow bench do show slight gains, but nowhere near 40 percent. The reduced stem also contributes to lighter valve weight.

While any street engine will perform fine with stock valves, if better valve life and more power are part of the plan, you might want to give my company, Lingenfelter Performance, or any of the many valve manufacturers, a call to discuss your needs. With so many size, shape and material opportunities, it's best to get a knowledgeable opinion before you buy.

VALVE SPRINGS

Even a street performance engine can push the limits of production parts,

especially when it comes to rpm and valve springs. The stock small-block Chevy valve spring works wonderfully for mild street use. However, its 80-pound seat pressure at its 1.70-inch installed height limits its performance, especially when asked to spin past 6000 rpm. The ability of the valve spring is further challenged by higher ratio 1.6:1 rocker arms and the latest high

The spring on the left is a stock diameter (1.250-inch) small-block spring with its flat wire damper. The spring on the right is a larger diameter (1.44-inch) spring for a small-block with a second round-wire inner spring. These larger springs are used for more radical roller cams.

Large valve springs on a small-block also make things tight for the head bolts. In this engine, the center head bolt washer was cut on both sides and a high quality Allen bolt was used to clear the two valve springs.

the main spring augmented with a flat-wire damper. In order to make room for this second, round wire spring, dual springs are larger diameter springs. While most aftermarket heads are designed to accommodate larger diameter springs, production heads are not. There is room on many stock heads for larger diameter springs. But if these killer springs are required, you're probably better off with an aftermarket head with greater flow potential to justify the use of the higher rpm levels these larger springs are designed to control. If you insist on machining stock heads for larger springs, use a .030-inch shim below the cutter to prevent intersecting the water jacket around the outside radius of the cutter.

Production small-block springs measure 1.240 inches in outside diameter. Within this small diameter valve spring dimension, many valvetrain companies have created stronger valve springs

initially to satisfy the oval track market that dictated the use of iron heads and "stock" springs. For example, Comp Cams offers three valve springs from 80 pound seat pressures up to 110 pounds at the same 1.700-inch checking height. At .450-inch valve lift, this stronger spring checks out at 285 pounds of pressure compared to a stock spring's 200 pound pressure. This spring is slightly larger at 1.250-inch O.D. but that is still small enough to fit within a stock spring pocket.

This is usually more than sufficient for virtually any small-block still using stock diameter springs. The best place to look for a spring recommendation is with the cam manufacturer. Each cam company usually lists a recommended valve spring that will complement the camshaft lobe profile. Roller cams require a higher spring rate than most flat-tappet cams since the roller's valve acceleration rates

are so much higher. Generally, the best combination for a roller is a larger diameter dual spring. However, the new generation of mild hydraulic roller cams will work fine with a good stock diameter spring. For the small-block, larger valve springs are usually from 1.380 to 1.540 inches O.D. Most street cams limit the valve spring diameter to around 1.440 inches. One major consideration with larger springs is that clearanced roller rocker arms are required. The best plan when making major cam and valvetrain changes is to rely on the cam company to recommend a complete, matching valvetrain package. This removes the guesswork from working with mismatched parts.

This emphasis on springs extends to maintenance as well. When an engine will no longer rev to its previous peak power point, weak valve springs are often the culprit. When rebuilding an engine,

These performance retainers are slightly lighter than factory retainers, but just as strong. Remember that larger diameter springs also require larger diameter retainers. When rebuilding any engine, new machined keepers are good insurance against dropping a valve.

it's always a good idea to replace the valve springs. If freshening an engine that is otherwise in good shape, check the spring pressures. Pressures that vary or are more than 10 percent lower than published specs should be replaced. This is inexpensive insurance, especially if you spin the engine higher. A strong indicator of damaged springs is valve lash that is difficult to maintain or a spring that is noticeably shorter than the rest when the engine is disassembled.

Retainers & Keepers

The valvetrain retainer's job is simple. Prevent the valve spring from shooting off the valve. The retainer works in concert with the valve keepers or locks to secure the retainer on the valve stem. All production small-block retainers are stamped steel for durability, as are the keepers. Performance valve springs typically will work well with stock retainers, but for durability it's always best to match the valve springs with

matching sets of retainers and keepers since the spring must fit the retainer properly.

For mild street engines, stamped keepers are inexpensive and completely reliable. But if you are stepping up to a bigger cam and stiffer valve springs, it's best to make the jump to machined valve locks to add that extra bit of reliability. There's much talk in the performance industry about 10 degree versus 7 degree locks. For all but the most killer street engine, the standard 7 degree locks are more than sufficient.

Reducing Weight—If rpm over 6000 is a part of your performance equation, weight plays a key role in valvetrain dynamics. For example, Chevy opted to dump the tin oil shedder on top of the 350 H.O. aluminum head valve springs to save weight while also lightening the stock steel retainers as a hedge against valve float. Any weight added to the valve, spring or retainer becomes a force that resists changing

direction at high rpm. This makes the valve spring's job that much harder at these elevated rpm levels. As noted below, valve float is a serious problem and not necessarily limited to high rpm race engines.

One option that may seem a bit exotic is a set of titanium retainers. Comp Cams, Manley and others offer these high strength retainers that can shave a few precious grams off the end of the valve spring, which is enough to add a few hundred rpm to what engineers call "valve toss rpm," which is the speed in which the engine goes into valve float. Since titanium retainers are lighter, I think these retainers can add perhaps an extra 100 to 200 rpm safety margin, allowing the engine builder to avoid ultra-high pressure valve springs. The advantage here is that killer springs usually generate extra heat and take their toll on pushrods and rocker arms.

Valve Float

Valve float is a much-abused term for a set of circumstances in which the valve spring can no longer control the action of the valve. This is rarely a problem in stock type engines that never see 5000 rpm, but in a performance engine, it's possible your engine has experienced valve float even without your knowledge.

There are a number of factors that contribute to valve float but weak valve springs, a heavy valvetrain or excessive engine rpm are the top three culprits on the valve float Wanted poster. While it may be fun to watch that tach needle spin around to 7000 rpm, most street engines give up on the power curve far below this rpm. So all that thrilling rpm only contributes to destroying what perhaps were a decent set of valve springs—if you're lucky. If lady luck has frowned on you, valves hitting the pistons, bent pushrods or worse will likely be the result.

The reason a valve spring can no longer control a valve is a bit complicated. A

The stock guideplate used on the Corvette 350 H.O. aluminum head is not hardened, although some of the early 'Vette heads did come with hardened guideplates. On all others, the guideplates must be removed and replaced with standard hardened guideplates that won't wear when used with hardened pushrods with other than rail rockers. These non-hardened guideplates were used to keep the pushrods in place on the production line before the rail rockers were installed.

spring that can control a valve at 6000 rpm may not be able to close that same valve at 6100. The valve, in essence, becomes ballistic, remaining open when the spring should be pushing it closed. This can occur because the valve, spring and retainer combination weighs too much, the rocker ratio is high (creating excessive valve accelerations) or perhaps the camshaft design is a little too radical for the spring. Any combination of these and other factors can cause valve float.

Regardless of the circumstances, the result of valve float even if the valves don't hit the pistons is the valve springs are immediately and permanently damaged. Typically, during valve float, part of the spring is attempting to compress while another portion of the spring is try to expand. This superheats the spring, causing permanent damage and lost spring tension. The immediate

result of valve float is power dropping off radically.

Once this occurs, the valve springs are usually damaged so badly that an engine that once was easily capable of 6000 rpm will no longer be able to reach that rpm. There are other factors that can contribute to an engine that won't rpm, but if the engine has been over-revved recently and won't respond to tuning changes, then weak springs could be the cause.

The cure for valve float is some combination of reducing weight in the valve and/or retainer, increasing valve spring tension or changing to a less aggressive camshaft. Lightening the rocker arm does little and while higher valve spring tension helps, it causes other problems by over stressing the valvetrain. The proper fix is an intelligent approach to designing the valvetrain and a fanatical adherence to a realistic redline.

PUSHRODS & GUIDEPLATES

If stronger valve springs and a little more rpm are in your game plan, then stronger pushrods and guideplates should be included. The standard small-block Chevy pushrod is made of hardened 1010 steel tubing that works fine for most mild small-block Chevys. This was true up until the introduction of the rail rocker arm in '86. Most enthusiasts don't know that these pushrods can work in performance applications with pushrod guideplates. If you're not sure if your pushrods are hardened, gently drag a file across the pushrod. If material is easily removed, the pushrod is not hardened.

Adding guideplates and screw-in rocker arms are a common modification to production iron small-block heads. Stock Chevy iron heads press the rocker stud into the head. With high rpm or

Rocker arm options for the small-block (from left to right) include the stock stamped rocker, a Comp Cams roller-tipped rocker, an aluminum roller rocker and a Comp Cams stainless steel roller rocker. Stamped rockers are fine for mild street use. If rpm is in the plan, choose a good roller rocker over the roller-tipped piece.

bending, which is usually blamed on the pushrod. In almost every case, however, a bent pushrod is a clue to a deeper problem that often relates to geometry. With stock rockers, check the condition of the stud and/or rocker ball. Excessive rocker, ball or stud wear allows the rocker to slide sideways off the valve stem, bending the pushrod. Weak valve springs that allow the engine to enter valve float will also cause bent pushrods.

Perhaps it is a sign of the times, but even when it comes to pushrods, the classic small-block Chevy's interchange-ability is becoming more complicated. Beware that pushrods for the late model hydraulic roller cammed engines are significantly shorter to accommodate the taller factory hydraulic roller lifter. Of course this means they are not interchangeable with typical flat tappet lifter applications.

ROCKER ARMS

It used to be there were two types of small-block rocker arms. Stock stamped rockers were used on all street engines and exotic roller rockers were the domain of the race engine builder. Now there are several stamped rocker arms including

higher rate springs, it's possible to pull the stud right out of the head. Screw-in studs prevent this and offer a way to retain the guideplates.

Guideplates

There are basically two different styles of pushrod guideplates. The most common is the flat steel guideplate, while the other is the raised guideplate, where the slots that guide the pushrod are moved closer to the end of the pushrod. While either style guideplate can be used with either stock or roller rocker arms, usually the flat guideplates are used with roller rockers and the raised style is used with stock rocker arms. Both of these guideplates are offered for either 5/16 or 3/8-inch pushrods.

Pushrods

For higher performance applications where a superior pushrod is necessary, Chevy offers two 5/16-inch pushrods, one of 1010 mild steel offered in both stock length and .100-inch longer and the other a very strong 4340 steel pushrod offered only in a +.100-inch length. This later pushrod is really intended for race-

only applications and not necessary for even a stout small-block. Of course all of the aftermarket cam companies also offer premium quality pushrods. Comp Cams, for example, offers three different material pushrods similar to those offered by Chevy with more length options to customize your valvetrain geometry.

The classic problem with pushrods is

Beginning in '88, Chevy changed the stock stamped rocker to a rail or "guided" rocker (top) that uses two small rails to retain the rocker on the end of the valve. This rail rocker does not require a pushrod guideplate. Rail rockers need a taller valve stem height above the retainer for clearance. Never use a rail rocker on an earlier valvetrain.

the production "rail" rocker and several styles of roller rockers.

Stamped Rockers

The classic stamped rocker arm that differentiated the first 265 cid small-block Chevy from all previous engines of its day is really an amazing piece of engineering. The simple stamping rides on a ball that allows it to pivot on the rocker stud while imparting a 1.5:1 ratio to cam lift. For mild street engines, this piece works fine and remained unchanged from 1955 until 1988. In that year, Chevy engineers introduced the guided rocker arm that uses two small risers or rails that straddle the valve stem to keep the rocker centered over the valve stem. Previous small-blocks used holes drilled in the heads to maintain pushrod alignment.

Rail Rockers—With the advent of aluminum heads, Chevy employed the rail type rocker. This rail rocker requires a taller valve stem tip that protrudes further above the retainer. This is important since rail rockers cannot be used on older, shorter valve stem tip engines. Neither can non-rail stamped rockers or roller rockers be used on the '88 and later heads unless hardened guideplates are employed to properly align the pushrods (see "The Great 'Vette Head Hunt").

Among the rest of the small-block Chevy family, companies like Comp Cams, Crane and others also offer stock-appearing stamped rocker arms that increase the rocker ratio from 1.5 to 1.6:1. This .1 ratio increase typically increases valve lift around .030 inch. This also places more pressure on the rocker stud and ball. While this is an easy and inexpensive way to increase lift, generally this move does not dramatically increase power when used in conjunction with stock iron cylinder heads. This is because most stock or near-stock iron heads don't flow well above .400-inch valve lift.

THE GREAT 'VETTE HEAD HUNT

The late model Corvette and H.O. 350 aluminum cylinder head is quickly becoming a popular street head because of its inexpensive price, good flow potential and light weight. All the heads come with what appear to be typical hardened guideplates. The factory uses these plates to maintain alignment of the rail type rockers. In a performance application, adding any other type of rocker arm will require the addition of a set of hardened guideplates. I use the Chevrolet high-performance guideplate part number 14011051.

Another pitfall that even many cylinder head shops may not be aware of is that the early 1986 and early '87 126 casting number aluminum 'Vette head employed an exhaust valve rotator. This thicker rotator was compensated for by increasing the installed height of the valve by roughly .120 inch. Later 'Vette heads eliminate this rotator and it is acceptable to remove the rotator to cut the additional weight as long as the proper thickness of valve spring shims are used to re-establish the correct installed height. If the proper installed height is not established, the exhaust valve spring will have insufficient spring pressure which will allow the exhaust valve to float prematurely.

The early, first generation Corvette head (casting number 128) used a thick exhaust valve rotator. To compensate, the distance from the spring cup to the bottom of the retainer is .120 inch taller on the exhaust than the intake. I remove this rotator, installing sufficient shims underneath the exhaust spring to compensate for the removed rotator.

Roller Rockers

Roller rockers have become more affordable and popular but there are a few misconceptions about roller rockers that should be addressed. Perhaps the most misunderstood point is that the roller tip actually rolls across the valve stem tip. This is not true. Even though the roller will spin easily between your fingers, once loaded against valve spring pressure the roller merely slides across the valve stem.

Late model hydraulic roller cam engines use a different cam snout than earlier small-blocks. The cam on the left is a late model hydraulic roller cam while the cam on the right employs the earlier style cam mounting surface. You must match the appropriate cam gear and timing set with these different camshafts.

well to valve lift increases much beyond .420 inch since flow bench tests reveal that most stock iron heads max out at around that lift.

However, there are situations where experimenting with rocker ratios between the intake and exhaust lobes can be beneficial. Increasing rocker ratio does change the camshaft's duration slightly but the main gain is the added lift throughout the valve curve. One classic trick is to place 1.6:1 rockers on the exhaust side only. This improves the exhaust side relative to the intake. If power increases, it points out that either the exhaust port needs more work or that a longer duration exhaust lobe would help power. The next step would be to add 1.6 rockers to the intake to see if both 1.6 rockers help power. Trying just 1.6 rockers on the intake alone can also be tried, especially on heads with an exceptional exhaust port.

Remember that the rocker arm travels in an arc across the top of the valve stem as it opens the valve. The rocker arm

The advantage of a true roller rocker arm is that roller bearings in the trunion, or pivot, dramatically reduce friction compared to the stock ball and stud used on a stamped rocker arm. This is evidenced by the 10 to 15 degree reduction in oil temperature in an otherwise stock engine when true roller rockers replace stock rockers. For stock engines, stock rockers are proven performers. If you want to spend some money to update your valvetrain, step up to true roller rockers.

Rocker Ratio

Rocker arms multiply cam lobe lift by a ratio in order to increase valve lift. The stock small-block Chevy's rocker arm ratio is officially 1.5:1. However, stock, stamped steel rocker arms are notorious for being as short as 1.45:1, which can cost .015 inch of valve lift. High performance roller rockers are more accurate, and often deliver slightly more than their advertised ratio.

Conversely, increasing the stock ratio from 1.5 to 1.6:1 for example will increase valve lift usually by about .030

inch. Changing rocker ratio is the quickest way to determine if increasing valve lift will improve power. Often, stock Chevy cylinder heads don't respond

On the left is what is referred to as a link-belt timing chain gear set found on production engines. The set on the right is a roller chain and gear set. There are many different versions of a roller timing chain set so be careful when choosing.

I prefer to use a plastic cam button on aftermarket roller cam engines to prevent cam walk. Aluminum buttons are also available but they tend to gall the inside of the timing chain cover. I use only bronze thrust washers behind the cam gear. Those trick roller bearings sometimes fail and those tiny bearings end up destroying the oil pump!

moving through this arc will change the ratio slightly throughout the lift curve. All rocker arms suffer from this condition although stock stamped rockers are the worst offenders. Aftermarket roller rockers tend to compensate for this condition by increasing the ratio beyond the advertised ratio.

TIMING SETS

It should come as no surprise by now that there are several choices facing the small-block engine builder when it comes to timing sets. Decades ago, the old Chevy plastic-coated timing gear and its link-belt chain were the only choice. Racing pushed the envelope and soon the roller timing chain replaced the old Chevy gear as the high performance standard. The term "roller" in roller timing chains refers to round bearings used in the timing chain that roll across the teeth in the timing gear as opposed to the stock type link belt timing chain that creates more heat since there are no rollers.

Most premium roller timing chain sets also improve the quality of the gears with hardened steel rather than the cast iron found in stock type link belt sets. This is for added strength and to be compatible with the harder material used in the roller chain to reduce wear. There are a number of good timing chain sets available for the small-block from a number of different outlets, but I prefer to use the Speed-Pro 4100 sets for most applications. These hardened timing sets also require some type of bushing between the block and the gear. I have found that those trick roller bearings tend to fail and find their way into the oil pump where they quickly lock up the pump! Instead, I use a Longacre bronze shim that often requires little or no gear machining to place between the cam gear and the block.

Building performance engines is a matter of pushing the envelope. High output engines quickly identify the weak link in the chain. Perhaps nowhere else in the small-block Chevy is this point hammered home harder than in the valvetrain. This may be a repetitive point, but spending good money for quality parts, especially in the valvetrain, will pay off with not only excellent power, but increased durability as well. That means you can hammer your motor harder without fear of flying... parts. ■

14 INDUCTION

By now, you might begin to see that specific parts combinations are the key to a powerful small-block. If you're reading this book chronologically, you might want to go back and review Chapter 1 on *Basic Engine Theory*. Before we dive into all the trick parts and ideas in this chapter, it might be wise to go back and refresh those concepts.

With the theory from Chapter 1 fresh, remember that changing runner lengths and intake runner size is one of the best ways to affect the engine's power curve. We'll take a look at both carbureted and electronic fuel-injected manifolds and also give you some hard-earned inside hints on how selecting the proper intake manifold is critical to optimizing power from your combination.

Changing runner lengths and intake runner size is one of the best ways to affect an engine's power curve.

CARBURETED INTAKES

There are two basic styles of carbureted intake manifolds: single plane and dual plane. The two styles are different in both appearance and in performance. The dual plane intake is designed for low and mid-range power by using two distinct and separate plenum chambers to feed two different induction paths. These runners tend to be long and small in diameter. The combination of long, yet small diameter runners enhances low and mid-range power.

Dual Plane

This is the common design for the small-block Chevy production intake. Even the 1970 370-horse LT-1 came with an aluminum dual plane intake. The disadvantage of the dual plane is that since the runners tend to snake around each other in the manifold, their size must remain small. The two separate plenum chambers must also be small for packaging reasons. Therefore, the dual plane is limited in top end horsepower

compared to a single plane intake. To their credit, manifold designers have made great strides in dual plane intake design in the last few years. The best overall dual plane intakes currently are the Edelbrock Performer RPM and Holley Contender. These manifolds have proven themselves to create a broad and strong torque curve over a wide rpm band while sacrificing only a small amount of horsepower to the single planes at engine speeds above 5800 to 6000. Either intake would be an excellent choice for any

Carbureted intake manifolds come in two basic configurations for the small-block Chevy. The dual plane intake like the Edelbrock Performer RPM (left) uses two completely separate chambers that each feed four cylinders. The design creates longer intake runners that generate more low and mid-range power. The single plane intake like the Victor, Jr. (right) utilizes larger runners than the Performer RPM that are also much shorter. This combination pumps up the upper-end power.

When you consider that a NASCAR 358 cid small-block can make over 700 horsepower with only one 830 cfm and a Bow Tie intake, it makes you wonder why anyone would want or need a tunnel ram with two monster carburetors for a small-block. Tunnel rams for the street are generally a waste of time.

into the short runner category, usually as upper rpm, horsepower-type intakes that usually sacrifice low-end power.

There are a number of variations on the single plane theme. Among the most popular and best performing single planes are the Edelbrock Victor, Jr. and the Holley Strip dominator intakes. Both offer great power with a very broad power band. In fact, when compared to the two previously mentioned dual planes, there is a significant overlapping power band in the mid-range where both styles make similar power. In some combinations, it's possible to make more mid-range power with a Victor, Jr. than with a Performer RPM. It's only at the very bottom where the dual plane shines and at the very top at horsepower peak where the single planes are dominant. There are a number of other single planes offered by a host of manufacturers, including Brodix's new single plane designed to work with the Brodix heads that looks promising, but I have yet to test this intake.

Given allowances for gross oversimplification, dual plane intakes tend to work better with mild and medium performance small-blocks especially when combined with camshafts of duration at .050-inch tappet lift of less than 225 to 230 degrees. Single plane intakes then take over and work well with engines with good cylinder heads and camshafts with duration figures at .050-inch tappet lift of above 225 to 230 degrees, especially with roller cams. Obviously, there are exceptions to these applications. For example, a 383 with good heads and a roller or hydraulic cam of 230 degrees of duration would respond to an Edelbrock Performer RPM, especially if that engine was in a heavy street car equipped with an automatic with less than 3600 rpm stall speed. As stated in Chapter 1, the choice of an intake can be affected by the car itself in terms of weight, gear ratio and transmission.

street-driven small-block.

Single Plane

The single plane intake is distinctly different from the dual plane with its large plenum or open chamber directly beneath the carburetor. This plenum directly connects all of the intake manifold's eight runners. This design generally assigns all single plane intakes

There is more in the world of high performance carburetion than just Holleys. The Rochester QuadraJet (left) is an excellent street carburetor that will flow in excess of 750 cfm with a highly-tunable primary circuit. Edelbrock has recently added the Carter AFB market to their line of Performer carburetors (middle) while the Holley remains the standard. HPBooks offers excellent Holley and Rochester books.

CARBURETORS

Whole books have been written on the subject of carburetors, so we will be brief. The choice of an intake manifold is probably much more important than the carburetor you choose to bolt on top. Obviously, the carburetor must be of sufficient size to not become a restriction. The most important task for the carburetor is to generate a fuel curve that will be the ideal air/fuel ratio throughout the entire power band of the engine. For street engines, this means the carburetor will be asked to accurately meter fuel from idle to 6500 rpm, or higher. This is a daunting task for any carburetor.

Holley

For wide open throttle maximum power and simplicity, the Holley carburetor is still the king. It's hard to go wrong with a 750 cfm vacuum secondary or double pumper Holley. I prefer the double pumper because I feel the vacuum secondary Holleys tend to be overly rich when the secondaries open to compensate for the lack of a secondary squirter. The

key to a properly set up Holley is to establish an optimized fuel curve the engine wants which goes beyond jet changes to include booster, main well and air bleed modifications. These modifications are best left to professional carburetor tuners.

Q-Jets

This is not to suggest that the other carburetors on the market are inferior mix masters. I have won numerous national event titles in NHRA Super Stock using a QuadraJet carburetor. When the Q-Jet is assembled and tuned properly it will deliver outstanding street performance. It's unfortunate that the Q-Jet's perceived complexity baffles many enthusiasts. The Q-Jet's metering rod and jet arrangement combined with its sensitive primary metering circuit is what gives the carb its great street manners capable of excellent part-throttle fuel management. As a plus, few realize that the Chevy Q-Jets are rated at 750 cfm. This is more than enough airflow for upwards of 500 horsepower. A great feature is that secondary metering rod changes can be

accomplished in seconds by merely removing one small screw and lifting the hanger.

Carter

The Carter carburetors are also good choices and even share many attributes with the Q-jet such as a primary metering rod and jet arrangement and a secondary air valve in some applications. One nice thing about the Carter is that the primary metering rods can be quickly changed without removing the lid that's required on the Q-jet.

Carb Selection

For mild street engines, smaller cfm Holleys like the 0-1850 600 cfm vacuum secondary carb will work fine, but given a choice, the Q-jet is a better choice for performance and accurate part-throttle fuel metering. The Carter falls into this category also. However, once you move into the heavy-throttle, over-400 horsepower street engines, the 0-3310 750 vacuum secondary or the 0-4779 750 double pumper Holleys are both excellent choices.

The Q-Jet offers tremendous tuning capabilities in part throttle operation. The combination of primary metering jet, metering rod, power valve spring along with being able to adjust the height of the metering rods in the jets gives a tuner a wealth of options. For everyday street-driven engines, the Q-Jet can be a precise fuel mixing device.

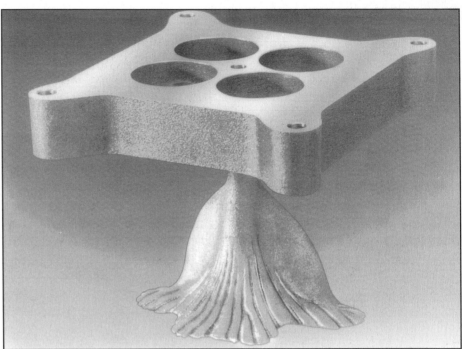

Carburetor spacers affect the speed of the air and fuel into the intake manifold. The most often used spacer is the open plenum spacer used on a single plane intake. Jean Dittmer has taken this idea one step further with the "Turtle." The Turtle is designed specifically for an individual intake manifold and direct air and fuel for improved mixture distribution to prevent lean cylinders. The Turtle is available through Brodix.

Carb Spacers

Carburetor spacers are a simple way to adjust the power curve on either single or dual plane intake manifolds. There are basically two types of carb spacers, open style or four-hole spacers. The open type spacer adds plenum to the existing intake manifold while the four-hole spacer extends each individual barrel of a four-barrel carburetor. Of the two, the open hole spacer is the more popular and these spacers are available in a range of heights depending upon how much plenum volume you wish to add.

Adding plenum volume to a single plane intake may seem overkill, especially for some of the larger plenum intakes. The real function of the spacer is adding height between the bottom of the carburetor and the floor of the intake manifold. At high engine speeds, the velocity of the air and fuel exiting the carburetor is very high. Since air is lighter than fuel (even finely atomized fuel), the air is able to make the 90-degree turn into the port but fuel will tend to maintain its straight path and strike the plenum chamber floor, turning into little rivers of fuel which then snake their way into various ports. This is one reason why maintaining consistent mixture distribution between cylinders with a carbureted engine is so difficult.

Adding a 1-inch spacer underneath the carburetor adds plenum volume which slows the velocity of the air/fuel mixture. This extra distance gives the mixture

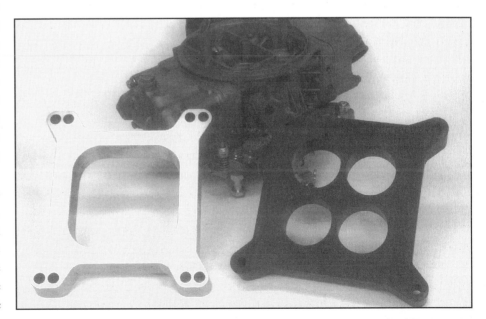

Carburetor spacers come in two basic styles, the open spacer (left) and the 4-hole spacer (right). You can use these spacers to fine-tune your engine's power and fuel curves.

CARBS VS. FUEL INJECTION

Old wives' tales and misinformation are common in any technical venture and high performance engines are no exception. Many enthusiasts mistakenly believe that there is some magic to electronic fuel injection that is worth instant horsepower. This is not the case. An engine's ability to generate power is simply the result of its airflow, combustion and exhaust capabilities. Intake manifolding directly affects both airflow and tuning and therefore can significantly change the engine's power band. Whether there is an advantage to how fuel is introduced, either from a fuel injector or by a carburetor, has little real effect on power as long as fuel distribution and droplet size are properly maintained.

Early in the development of performance variations on the TPI intake, I performed a very interesting dyno test to evaluate the power difference between a manifold outfitted with both a carburetor and then multi-point fuel injection. To ensure proper wet flow with the carburetor, I used a proven Victor, Jr. manifold. First, I tested the intake on a 383 small-block optimizing power with proper jetting. I then outfitted this same intake with eight, multi-point fuel injectors and bolted on a modified Holley carburetor as a throttle body. No other changes were made to the engine. Once the fuel curve was optimized for best power, the two power curves were compared.

As you might guess, the engine made virtually the same power with either system. What this means is that runner length and size is more important to the power curve than how fuel is introduced into the engine. Remember, this is just a comparison of wide open throttle power. With a long duration cam and big heads, an EFI system would realize better throttle response since it does not rely on a weak, part throttle signal to initiate fuel flow from a carburetor.

I configured this Victor Jr. to run either a carburetor or multi-point fuel injection. In comparison tests, there was no more than a 5 horsepower difference between the Holley carburetor and the EFI system.

more time to slow down, allowing the fuel to more easily make the turn into the port.

The net effect of this change is that adding the carb spacer tends to improve top-end horsepower while reducing low-speed torque below torque peak. Often, this horsepower gain can be an advantage if the power increase comes within the rpm range where the engine spends most of its time. If not, as is often the case, the horsepower gain is negatively offset by

the torque loss and the car will be slower in the quarter-mile.

Also keep in mind since an open carburetor spacer adds plenum volume, this also tends to reduce the signal to the carburetor. This usually results in a carburetor that will be slightly leaner than without the spacer in place.

Four-hole spacers can also be used to improve throttle response on street-driven engines. In fact, many drag racers use a four hole spacer directly below the

carburetor to maintain velocity, then use an open spacer to transition into the manifold. You might want to experiment with four-hole and open spacers to determine the affect these spacers have on your combination.

THE PROMISE OF EFI

If there is one new "gotta-have-it" piece in the high performance game today, it's probably electronic fuel injection (EFI). As is usually the case, it

There are two styles of electronic fuel injection, throttle body injection (TBI) and multi-point. TBI units offer few advantages over a well-tuned carburetor. The term throttle body should not be confused with the unit used to house the throttle blades for a multi-point unit. This is one of my billet throttle bodies on a multi-point injected small-block in a '94 Firebird that will run at Bonneville.

took Chevrolet to lead the way with its successful V8 electronic fuel injection system, the Tuned Port Injection system (TPI). You may recall that Chevy experimented with the twin throttle body crossram system in the '84 Corvette, but that was a stop-gap venture that should be avoided as a performance option.

EFI can generally be categorized as either throttle body or multi-point systems. Throttle body EFI uses what amounts to an electronic carburetor, using two and sometimes four large fuel injectors housed in a two- or four-barrel body. This replaces the carburetor and injects fuel directly into the intake manifold. Throttle body systems are most often used in truck and base engine factory applications since this type of EFI

is less expensive to build and can utilize intake manifolds similar to carbureted intakes. This is what is termed a "wet flow" EFI system where the intake manifold directs both fuel and air as in carburetion.

How EFI Works

Carburetors can be made to make excellent power. NHRA drag racing and NASCAR stock cars have proven that. But carburetors require dedicated modifications to create these specific fuel curves. Jetting changes, for example, move the entire fuel curve up or down but can do little to affect the curve in one specific point. All engines require their own individual fuel curve. While carburetors can be modified to come

close to this requirement and do a great job for their relative low cost, a carburetor cannot compare to EFI in the ability to make very minor and specific changes to optimize the power and driveability of any engine.

Throttle Body—There are a number of different forms of electronic fuel injection. The least expensive form of EFI is throttle body where fuel is injected at the throttle body. The Holley EFI system is an example of this, as are the factory EFI systems used on pickups and some low output car engines. Since the fuel is injected into the manifold, this system differs little from a carburetor other than there is more precise control over the fuel.

Multi-Point—The next level up from

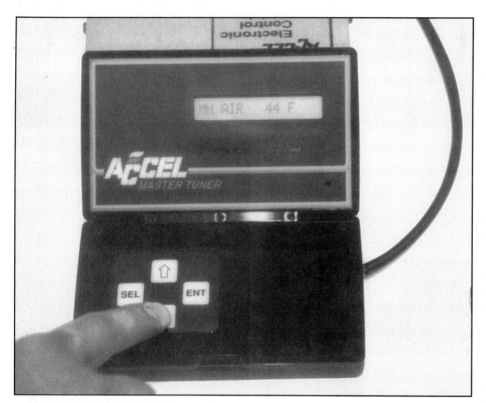

This is the new controller offered by **ACCEL** for its fuel injection system. This hand-held controller replaces the need for an expensive lap-top computer to make changes to the fuel or spark curve. This controller can be used on any **ACCEL DFI** system with an inexpensive computer and software upgrade.

throttle body is multi-point fuel injection. This is where hot rod style EFI really shines. Each individual cylinder retains a dedicated fuel injector that delivers fuel to only that cylinder. This generates more accurate mixture distribution between cylinders.

Speed Density—Within multi-point EFI there are a number of different styles. The most common aftermarket configuration is what is known as speed density. This is where the computer receives input from a number of sensors including a throttle position sensor (TPS) and a manifold absolute pressure (MAP) sensor. The MAP sensor tells the computer the load the engine is under. You can think of this sensor as an electronic vacuum gauge. If you've ever watched a vacuum gauge while driving a car then you know that as load increases by opening the throttle, manifold vacuum drops proportionately. The MAP sensor converts this load into a voltage signal. The computer plots this load reading on a

chart compared to rpm. This chart, or look-up table, has been programmed into the computer and instructs the computer to the amount of fuel to be injected by the fuel injectors. The amount of fuel to be injected is determined by the size of the injector, the length of time the injector is open (called the pulse width) and the amount of fuel pressure.

Mass Airflow—A second form of multi-point EFI is mass airflow where the engine actually measures the amount of air that the engine consumes through the use of a mass airflow (MAF) sensor. The computer then plots this information along with input from the MAP, TPS and other sensors to determine the amount of fuel to be delivered to the engine. While more accurate than speed density, the (MAF) also increases the price and complexity of the system.

Sequential EFI—Most factory and aftermarket EFI systems utilize what is called batch fire. In order to maintain precise control over the fuel delivered while keeping the electronics simple, most EFI systems fire the injectors once for every engine revolution, regardless of whether the intake valve is open or closed. A more accurate system called sequential EFI is equipped with a sensor that locates top dead center (TDC) of cylinder number one. This sensor informs the computer to the position of number one cylinder so that the injectors

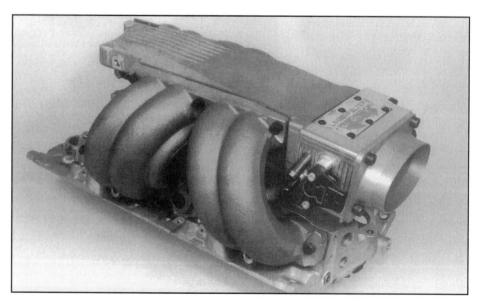

I have designed a number of pieces to upgrade the stock Chevy TPI for more power. This particular system uses my larger base manifold with High Flow TPI runners, a mildly ported stock plenum and a 1000 cfm billet aluminum throttle body with 58 mm throttle bores. This manifold makes astounding torque when coupled with my ported 'Vette heads and a matching cam.

Swapping a late model TPI system into an earlier hot rod is becoming more commonplace every day. The easiest way to accomplish this is with an ACCEL stand-alone computer. While a stock GM computer could be used, it's far easier to create a custom fuel and spark curve using the ACCEL system.

can be timed to fire just before the intake valve opens. This system is more complex, but does allow more precise control over the fuel.

Sequential EFI systems also allow the end-user to make slight changes to the amount of fuel delivered to individual cylinders to compensate for the flow differentials of either the intake port or intake manifold ducting. The late-model Camaro LT1 factory fuel injection system is a sequential EFI design while the older TPI systems were batch fire. ACCEL/DFI also offers an up-scale sequential EFI system that allows the engine builder to program fuel for each individual cylinder.

As you can see, there are many variations to the EFI game. Factory systems are efficient, but not easily adaptable to modified engines since their computers are not easily programmable.

For hot rodded engines, the best solution is a stand-alone, aftermarket EFI system such as those sold by ACCEL/DFI, Haltech, Electromotive and others or a piggy-back system such as that available from ACCEL/DFI. These systems have been proven through constant updating and most use either GM or Ford sensors that make repairs quick and easy.

TPI Manifold

The winner in the multi-point EFI game is the TPI system such as the one used on the Camaro and Corvette. A multi-point EFI gets its name from its multiple fuel injectors, usually one per cylinder. Highly accurate mixture distribution is created by dedicating a fuel injector for each cylinder. This reduces the problems of mixture distribution which is a common, although not serious, problem with both throttle body and

carbureted manifolds.

Design—Since a multi-point EFI manifold is a dry-flow design because it transports no wet fuel, it gives the intake manifold designer much more freedom. The original TPI intake manifold took advantage of this design freedom to increase runner length to create mid-range torque. An interesting aside to this story is that there is no difference in size between the 305 and 350 TPI factory manifolds. This is because the TPI was originally designed to increase the 305's torque, since misguided planners at design time had dropped the 350 from future Chevrolet plans! Thankfully, Chevy discovered that performance wasn't dead and quickly rushed the 350 back into the production plans. Unfortunately, Chevy was forced to use the 305 TPI manifold since no other performance intake existed. This explains

In this one photo is the entire range of intake possibilities for the small-block Chevy! On the far left is the new LT1 intake, then a carbureted dual plane, carbureted single plane and a tuned-up TPI on the far right. Our SuperRam is in the middle.

why the stock 350 TPI engine tends to give up at around 4800 rpm as opposed to a higher rpm point, because the intake was originally sized for an engine 45 cubic inches smaller!

This long runner length creates tremendous torque in the streetable rpm range between 2000 and 4000 rpm. The disadvantage is, as we've mentioned, the small, long runners are not conducive to high rpm power. Bolted to a 350, the stock manifold gives up way too early. Bolting a stock TPI intake to a larger 383 is like bolting on a 283 two-barrel intake and carb, giving new meaning to the term "induction-limited."

LT1 Manifold

The latest version production multi-point intake is the LT1 design. A number of non-engine design criteria affected the creation of this intake manifold. It is radically shorter than the previous TPI manifold, a requirement dictated somewhat by the sloping windshield of the fourth generation Camaros and Firebirds. This height limitation forced the manifold designers to build a short intake. In fact, the LT1's runners are shorter than an Edelbrock Victor, Jr. intake!

This shorter runner length creates a much broader power curve with a higher rpm horsepower peak. This shorter length surprisingly doesn't kill bottom end torque as much as originally thought, although the LT1 does not make as much mid-range torque as the longer runner

length L-98 TPI engine. However, my tests have shown that this shorter runner length exhibits poor fuel distribution during low engine speeds with high manifold vacuum when combined with higher overlap camshafts. This is evident at low rpm, part-throttle where the engine will surge lightly as intake pulsing affects cylinder filling of adjacent cylinders. This is only a mild driveability problem and one that only affects part-throttle cruise conditions. The addition of our Lingenfelter SuperRam on the LT1 improves this problem.

It's apparent that new LT1 intake was seriously limited in its design by the new Camaro and Firebird styling to the extent that Chevrolet engine designers were not allowed the freedom to create the

124

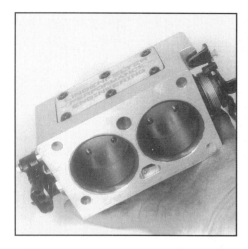

Lingenfelter Performance Engineering offers a billet aluminum 1000 cfm throttle body that starts showing a significant power improvement on engines making over 350 horsepower.

If the SuperRam looks like a shorter version of a High Flow TPI with a larger plenum, that's exactly what it is. We found that shortening the runners raised the peak horsepower rpm point with a sight increase in power between 3500 and 4000 rpm.

This is a Lingenfelter SuperRam stuffed in the engine compartment of a 1994 Pontiac Trans-Am. I had to trim the plenum height by 5/8 inch to clear the cowl. With this bigger manifold, it's a tight squeeze but worth the effort for the increase in torque it gives the LT1.

manifold they perhaps would have preferred. The aftermarket is not so limited, and as a result has come up with a couple of different manifold versions for the multi-point EFI small-block.

AFTERMARKET EFI MANIFOLDS

The different intake versions are perhaps best exemplified by my signature series manifolds available through ACCEL. I first began modifying the original TPI with larger diameter High Flow TPI Runners and a ported base. This worked well enough to warrant a larger port base to match the larger runners. There are also small gains to be made by mildly porting the stock plenum and adding my larger 1000 cfm throttle body in high horsepower applications.

During the TPI development process, I began experimenting with shorter runners in an attempt to improve top end power. Even with the larger runner diameters, the length of the runners prevented even a healthy small-block from making power above 5000 rpm, especially on larger small-blocks like the 383. Shortening the length of the runner by roughly 25 percent allowed the resonant pressure wave (which produces the ram effect that increases power at certain engine speeds) to occur at a higher rpm.

SuperRam

Our early experimentation lead to the development of the SuperRam intake. We cut down the runners and then constructed a large plenum box to connect all the runners to the stock TPI location throttle body. The large plenum's height is also designed to provide a sufficiently gentle radius to allow the air to "turn the corner" as it enters each of the eight runners.

Shortening the overall intake runner

There could be times when using a single plane intake with EFI would be advantageous. ACCEL has created this new intake that accepts electronic fuel injectors combined with one of my trick new throttle bodies. In ultra-high output systems, this could be combined with an ACCEL sequential fuel system that can individually modify the fuel curve to each cylinder.

Multi-point fuel injection for the small-block uses 8 individual fuel injectors, one for each cylinder. The injectors are rated in pounds-per-hour of fuel delivery. The more horsepower the engine makes, the more fuel is required, necessitating larger injectors.

length shifts the torque curve. Our dyno testing compared a 383 outfitted with a SuperRam base and High Flow TPI Runners to the same base with the SuperRam runner and plenum kit. Torque from 1600 to 3000 was virtually the same (the SuperRam was slightly down compared to the longer runners), at 3500 rpm the SuperRam pulled even and then from 4000 rpm up torque increased dramatically. In fact, at 5800 rpm, the SuperRam was up over 65 horsepower! This testing supports the theory that runner length will have an effect on the power curve since shortening the runners should increase top-end power at some sacrifice of low-speed torque.

Horsepower—Given the relatively long runner length of this manifold, some mistakenly think that the SuperRam is not capable of stout horsepower. Since creating this intake, I have built an 11:1 compression, roller cam 420 cid small-block using the SuperRam that made 525 horsepower at only 5500 rpm while

thumping out 540 lbs-ft. of torque at 4500 rpm. This same engine also grunted over 400 lbs-ft. of torque at 1600 rpm! I've also built a 406 that cranked out an astounding 585 horsepower and 585 lbs-ft. of torque with an excellent set of 18 degree Chevy heads using this same SuperRam intake. Both of these engines owe a portion of their power to porting of the heads, intake, runners and plenum. Our testing has shown that intake porting alone can be worth as much as 15 to 20 horsepower on a stout engine. I believe that this cylinder head combination on a 420 could easily make 600 horsepower with the SuperRam! Obviously, this intake can rock 'n roll.

Installation

The SuperRam runners and plenum are available separately if you are upgrading from an existing EFI system, or as an entire SuperRam package that includes the complete ACCEL electronics, wiring and fuel pump hardware necessary to

create a stand-alone electronic fuel injection system in virtually any vehicle. This installation does require some specific EFI installation knowledge and ability. It's beyond the scope of this book to detail all the steps necessary to make this conversion, but ACCEL has created a nationwide system of Electronic Management Installation Centers (EMIC) dealers who can install the system for you. You can find out more about the EMIC dealer nearest you by calling 800/992-2235. There are over 40 installation centers nationwide and two in Canada.

Performance Applications—The new LT1 intake can also be applied to a performance combination with a few minor porting modifications. Since this intake is extremely short, it gives up torque in the mid-range compared to the SuperRam, although its power curve is very flat. Using these two different intakes, we have created a number of emissions-legal performance packages primarily for late-model LT1 Corvette

The Chevy LT1 intake is an extremely short runner intake whose runners are actually shorter than an Edelbrock Victor Jr. intake!

and Camaro engines using this intake combined with our own camshaft and ported LT1 heads. The short runner intake works best with the 6-speed's close gear spacing. Conversely, the stronger mid-range power and wider power band of the SuperRam performs excellently with the automatic, although we have successfully used the SuperRam with the 6-speed when brutal street torque is the goal.

As mentioned in the aluminum cylinder head chapter, the new LT1 uses a different coolant path than earlier small-blocks, so there is no bolt-on interchangeability between earlier small-blocks and the LT1. This also requires changes to the SuperRam intake to allow it to be installed on the LT1 engine. I am currently working on a new SuperRam base manifold that will fit the SuperRam

directly to the new LT1 without modifications. In addition, 5/8 inch of height must be trimmed out of the current SuperRam plenum in order to clear the windshield cowl on the new generation Camaro/Firebird. This costs around 10 horsepower at peak power because of the tighter radius the air must negotiate entering each inlet runner and the reduced plenum volume.

MANIFOLD DESTINIES

As you can see, there are several intake tuning factors that remain constant even between carbureted and fuel injected engines. Intake runner size is perhaps the most important. Small runners generate great low speed torque but tend to limit top end power. Runners that are too large will be lazy at street engine speeds and

contribute to exhaust dilution/reversion in the intake. Short-length runners tend to emphasize top-end power while longer runners increase mid-range torque at some sacrifice of top-end rpm power potential. This is about as concise as you can get yet there are plenty of questions still to be answered. The future of induction tuning will address all of these factors plus a few dozen more. How about a variable-length, variable-diameter intake that could be long and small for good low-end power and short and fat for top-end power? If this sounds too much like Buck Rogers, it's already happened in Formula 1. Perhaps some enterprising hot rodder will someday build as practical and affordable an intake for the small-block Chevy! Think about it. ■

15 EXHAUST SYSTEMS

It's an interesting phenomenon that hot rodders generally overlook the exhaust as an avenue for increased performance. The reality is that the exhaust system is absolutely critical to engine performance. Even the lowliest 2-barrel 283 can benefit from improving the exhaust system, even if no other changes are made to the engine. I have found through testing that the very first modification made to a stock small-block should always be to improve the exhaust system.

This is almost universally true because factory exhaust systems are almost always restrictive, increasing exhaust back pressure in exchange for a quieter exhaust note. In the past few years, especially with the TPI Camaros and the fourth-generation LT1 Camaros and Firebirds, the exhaust systems have improved even with catalytic converters. In fact, our testing shows that the new LT1 catalytic converter costs only about 10 horsepower on a 400 horsepower engine, which is a small price to pay for clean air. However, these are the exceptions rather than the rule. In any performance combination, the exhaust system is the one area that cannot be slighted if optimal power is to be realized.

EXHAUST SYSTEM TUNING

The next major performance gains for

Exhaust systems are where you'll find your next major performance gains. When it comes to headers, size and length are key considerations.

high performance street engines will come as a result of exhaust system tuning. Currently, significant strides have been made in creating high performance exhaust systems for street cars that not only mute the exhaust note but do so at very little cost to horsepower. In fact, it is a little-known fact that adding a well-designed muffler and exhaust system to a powerful small-block will generally increase torque while slightly decreasing peak horsepower. Drag strip testing then reveals that this torque increase usually improves acceleration while the top-end

horsepower loss results mainly in a small loss in trap speed. The bottom line is that running the car with a well-designed exhaust system and mufflers will often improve the e.t. while losing only a slight amount of mph trap speed.

We'll investigate several different factors that contribute to exhaust system efficiency in this chapter. Most of this will be directly applicable to street engines rather than pure race engines. The best place to start is at the beginning with headers.

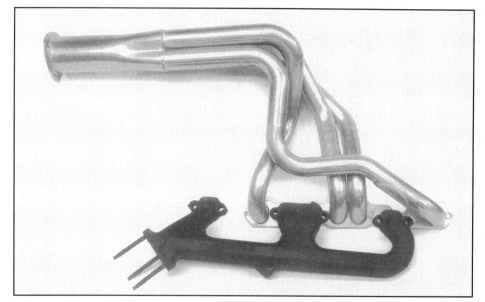

Headers offer a significant performance advantage over those ugly and heavy cast iron manifolds. The more horsepower the engine has potential to make, the more headers are worth. All headers are mandrel-bent which maintains a smooth inside diameter for improved flow. Usually, longer headers tend to improve low-speed torque.

Our dyno testing has confirmed that the late-model LT1 and LT5 catalytic converters are actually very efficient. On a stock '93 LT1 350 engine, removing the cat improves the top-end power only about 4 to 5 horsepower. The stock catalytic converter on our 500 horsepower killer LT5 engines reduces peak horsepower only about 12 to 14 horsepower. This illustrates the efficiency of these catalytic converters.

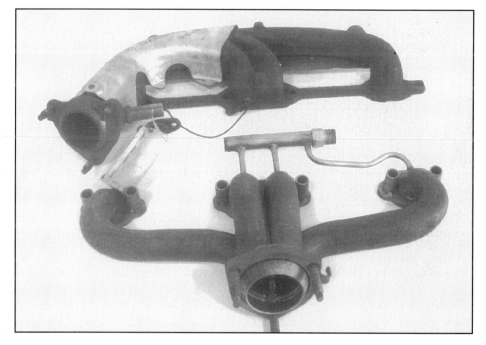

Chevrolet realized that a better exhaust system was worth power, which is why the Corvette was given stainless tubing headers (bottom). While the shorty, 1 1/2-inch tubes may look wimpy, I use them in all of my performance applications because they cost little in the way of power. The cast iron manifolds are used on the new LT1 Camaro and Firebird.

Headers

As with any good air pump, you cannot make more power if you cannot exhaust the spent exhaust gases from the combustion chamber. Exhaust manifolds do an acceptable job of this at low engine air flow, but suffer from significant flow losses at higher rpm. Since we know that higher revs can also contribute to greater power, assuming the engine is tuned for that power, then the exhaust system plays an increasingly important role. Headers typically do a much better job of efficiently evacuating the cylinder of spent exhaust gas at these high engine speeds mainly because they reduce the flow losses of a cast iron manifold. A secondary benefit is that these longer exhaust pipes contribute to helping tune the arrival of the scavenging or reflected wave during valve overlap to improve cylinder filling.

The reason typical cast iron exhaust manifolds do such a poor job of contributing to horsepower is that these manifolds restrict exhaust flow, which increases back pressure, and also because the manifold is too short to assist in wave tuning.

Header Size—Typical exhaust headers for the small-block Chevy are the four-into-one design with lengths between 32 and 36 inches. The size of the headers and the length of pipe both contribute, but the size of the pipes are more important than the length.

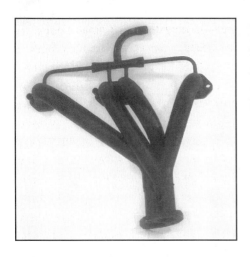

Tri-Y headers typically increase mid-range power slightly over 4-tube headers. The main advantage to Tri-Y's is the fewer tubes are easier to squeeze into tight engine compartments and sacrifice little peak power.

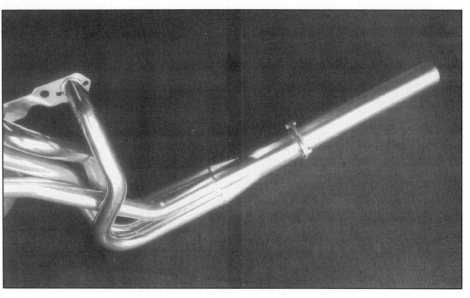

For drag strip open header applications, adding collector length can improve bottom end torque. You will need to experiment with your combination, but adding roughly 8 to 10 inches of collector length can boost low-end torque.

Equal Length—There is also more than a little discussion about the importance of equal-length headers. Most authorities agree that equal-length headers are important, but mainly when applied to finely-tuned race engines. I feel that while equal-length headers may be of some advantage, they are not worth the much higher price these headers usually demand. Current street headers from all the different header companies can vary primary header tube (the tube from the exhaust port to the collector) length by as much as two to four inches.

Pipe Diameter—Primary pipe diameter is far more important to ultimate engine power than pipe length. Header pipe size is a compromise between exhaust gas velocity and pipe diameter sufficient to handle the mass flow of the exhaust gas. Smaller header pipes increase gas speed at low rpm which efficiently evacuate the chamber at these low speeds, contributing to improved low-speed torque. Unfortunately, these same smaller pipes become a restriction at higher engine speeds.

Conversely, a large primary header pipe tends to be "lazy" at lower engine speeds with lower exhaust gas speeds at low and mid-range engine speeds, which does

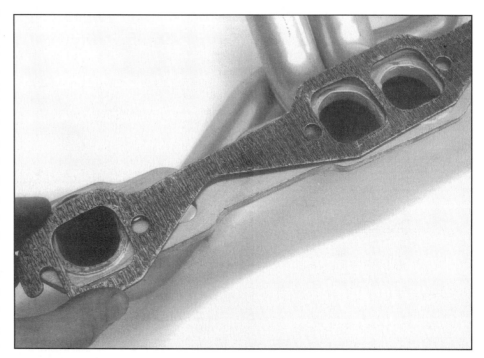

One of the biggest headaches with headers is leaking gaskets. Fel-Pro has solved that problem with its steel core construction flange and collector gaskets. The high temperature gasket material is backed up by a steel core foundation that maintains the integrity of the gasket far longer than typical paper style gaskets. Fel-Pro offers these gaskets in 9 different exhaust port configurations for the small-block.

allow a certain portion of the exhaust gas to remain in the chamber as the exhaust valve closes. Since this residual exhaust gas cannot burn twice, it contaminates the incoming fresh air/fuel charge, lowering the cylinder pressure during the next combustion cycle, resulting in less power.

Daily-driven street performance engines tend to spend a majority of their time at low and mid-range engine speeds. Choosing a header for this application means staying on the small side of header pipe size in order to enhance low- and mid-range engine performance. Matching

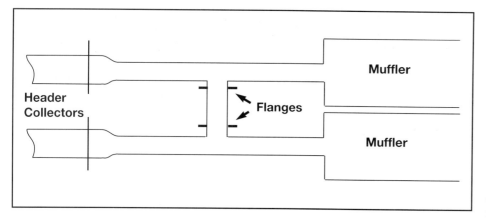

Placing a cross-over or H-pipe in the exhaust behind the collectors but in front of the mufflers is a quick way to both dampen exhaust noise and increase torque.

LINGENFELTER EXHAUST COMPONENTS

We have created six different header combinations for the late model Camaro/Firebird and Corvettes. These headers are both four-into-one and Tri-Y design and are offered for high output, off-road applications only since the headers have not been certified for use in emissions-controlled vehicles even though they do make allowances to install all the factory emission control devices such as air pump fittings and oxygen sensors. The Y pipe used to connect the headers to the converter is also included and the headers are coated with a ceramic coating to prevent corrosion.

HEADERS

APPLICATION	PRIMARY PIPE SIZE (inches)	PART NUMBER
Corvette 1984-'85	1-5/8	74410
Corvette 1986-'91 Coupe	1-5/8	74411
Corvette 1986-'91 Convertible	1-5/8	74412
Corvette 1986-'91	1-3/4	74413
Camaro/Firebird 1982-'91 (single catalytic converter)	1-5/8, 1-3/4, 1-7/8	74400
Camaro/Firebird 1982-'92 (dual catalytic converters)	1-5/8, 1-3/4, 1-7/8	74401

Note: The Camaro/Firebird headers are a stepped Tri-Y design while the Corvette headers are the traditional 4-into-1 style.

this with a short duration camshaft and dual plane intake results in a responsive, torquey and fun to drive small-block. To help in choosing a header size, we've included a Header Size chart that looks at displacement and peak horsepower as its reference points. This is a very generalized chart, but will help in selecting the proper header pipe diameter.

Tri-Y—An alternative to the classic four-into-one header is the Tri-Y design. Tri-Y's get their name from how they join the four separate header tubes. Four-into-one headers join all four primary tubes at the collector while Tri-Y headers join two primary pipes together ahead of the collector, giving the header the appearance of forming three Y unions. This design shortens the primary pipe length and adds a secondary pipe before the exhaust reaches the collector. The general consensus among engine builders is this design broadens the torque curve while four-into-one headers tend to be more peaky. The primary builder of Tri-Y headers is Doug Thorley Headers.

Lingenfelter—We have also constructed a number of headers specifically for the 1982-'91 Camaro and Firebird and also for the '84-'91 Corvettes to complement our other emissions-legal components offered through Lingenfelter Performance Engineering. The headers for the Corvettes utilize the four-into-one design and are offered in both 1-5/8- and 1-3/4 inch primary pipe diameters. The smaller headers are designed for engines making 380 horsepower or less, with the larger headers intended for the higher horsepower engines. The F-car Camaro and Firebird headers are similarly sized but are built in the Tri-Y design due to chassis restrictions. All of our headers are constructed of 14-gauge, mandrel-bent mild steel tubing with a ceramic coating for corrosion protection. All emissions-required connections such as the air tubes and oxygen sensor bosses are included on every set of headers. Doug Thorley also builds excellent Tri-Y headers for a wide variety of other car and truck applications.

The Whole System Approach

Building a high performance exhaust system doesn't stop with adding a set of headers. In fact, it's possible to gain little torque or horsepower from adding a set of headers if the exhaust system and mufflers are restrictive. The result is to merely move the cork further

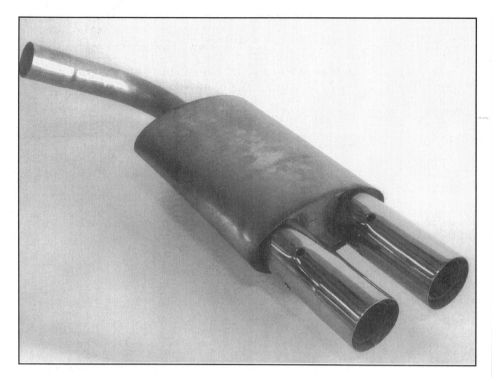

I have found that the Borla mufflers produce the best overall power for our engine combinations while maintaining their great appearance. All Borla mufflers are produced out of T-304 stainless steel and have a lifetime warranty. If you are looking for the ultimate performance muffler, this is it.

HEADER SIZE CHART

The following chart should be used as a general guide to header primary pipe size selection based on engine displacement and horsepower. There will be exceptions to some of these applications, but generally these pipe sizes are close. It may be difficult, if not impossible to find 1-1/2-inch primary pipe headers for a small-block Chevy. In this case, the 1-5/8-inch headers will work. If your application falls somewhere between these suggestions, choosing the smaller of the two header pipe sizes would tend to increase torque while the larger header would tend to increase top-end power. Also remember to rate your engine's horsepower conservatively. There are probably very few 500 horsepower 305 cubic-inch engines. Please refer to the small-block dyno tests in Chapter 18 for horsepower levels. A 9:1 350 with production heads and a mild hydraulic cam probably makes less than 300 horsepower.

ENGINE DISPLACEMENT

	305	350	383	406
HP				
300	1-1/2	1-5/8	1-5/8	1-5/8
350	1-1/2	1-5/8	1-5/8	1-3/4
400	1-5/8	1-5/8	1-3/4	1-3/4
450	1-3/4	1-3/4	1-3/4	1-3/4
500	1-7/8	1-7/8	1-7/8	1-7/8

downstream. The most efficient exhaust system is a combination of pipes large enough to not restrict the exhaust while small enough to provide sufficient exhaust gas speed. Combine this with pipes that keep the number of bends to a minimum since each individual bend can contribute to flow losses. Even more important is how these bends are produced. Mandrel-bent tubing is the most efficient since the tubing bender uses a solid steel mandrel inside the tubing that prevents the crimps that you see on production exhaust tubing. These crimps in the tubing reduce the inside cross-sectional area of the tubing which increases flow losses.

H-Pipe—The typical exhaust design is a pair of exhaust pipes leading to a pair of high performance mufflers. One addition ahead of the mufflers that has been proven to improve both power and reduce noise is an H-pipe. The H-pipe (also called a balance tube) is nothing more than an additional pipe placed between the two lead-down exhaust pipes. The H-pipe tends to increase

volume in the exhaust system ahead of the mufflers. Our testing has shown that a large diameter H-pipe tends to increase torque at or below peak torque while not detracting from top end horsepower. You can look at the H-pipe as increasing the length of the header collector, which has a similar effect on power. Additionally, by connecting the two separate exhaust pipes, the H-pipe also dampens noise. Placement and design of the H-pipe is not very critical, as long as the pipe is placed between the header collectors and the mufflers. Recent testing of H-pipes also suggests that a larger H-pipe seems to benefit torque. If you place flanges on both sides of the H-pipe, it also makes it easier to install and remove, which reduces the hassles when it comes time to pull the transmission.

Mufflers

Mufflers have always been viewed as a necessary evil by hot rodders. Engines are inherently noisy rascals and especially for daily-driven street cars, an overly loud car is rarely a good idea despite

adolescent fantasies to the contrary. There are generally three approaches to muffling engine exhaust note.

Restriction—Most low performance mufflers use restriction to reduce noise. By requiring the exhaust to travel through a maze of tight turns and slowing the exhaust gas down, noise is dissipated and the exhaust note is dramatically quieter.

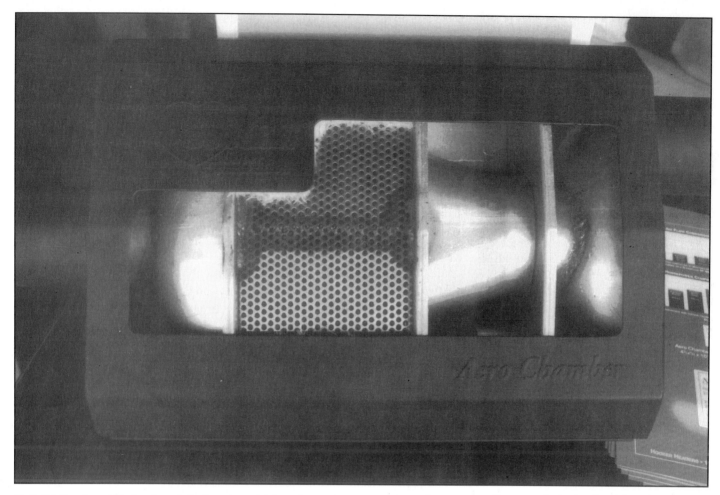

Both Edelbrock and Hooker have introduced new high performance street muffler designs lately. This is the new Hooker muffler. While I have yet to test them, they appear to all perform similarly to the Flowmaster and Walker Super Turbo mufflers.

Unfortunately, restriction is always accompanied with decreased flow, resulting in back pressure.

Back pressure is the increase in exhaust system pressure that negatively affects exhaust flow. Many restrictive stock exhaust systems can create excessively high back pressures that can exceed five to six psi. This generally results in a pronounced hissing sound coming from the exhaust system under wide open throttle. This is a classic example of excessive exhaust back pressure. Merely adding a high performance exhaust to a vehicle with this problem will result in a dramatic performance improvement. The sidebar "Pressure Test" outlines how you can measure exhaust system back pressure easily and inexpensively.

Absorption Method—The second and most popular way to muffle sound is

with absorption. The glasspack muffler popular in the 1960's is an example of an absorption muffler. A tube usually the same size as the exhaust pipe is punched with louvers and then wrapped with an absorptive, heat resistive material like fiberglass matting and then covered with an outside pipe. As the sound waves travel through the muffler, the louvers allow the sound waves to be absorbed in the fiberglass matting. Looking down a straight-through glasspack muffler, it would appear to be free-flowing. Unfortunately, the louvers that protrude into the flow path also create back pressure, which is why glasspacks are never a good performance muffler and are a rare sight on anything except on a few lead-filled customs.

Noise Cancellation—The third technique for muffling noise is reflective

or noise cancellation techniques. At some point, you may have noticed how two waves traveling directly at each other in a water tank will cancel each other out. This same effect can occur with exhaust pressure waves if they are directed at each other. The Flowmaster muffler utilizes this technique to reduce exhaust noise. One advantage to noise cancellation is that it creates little back pressure, although changing the direction of the exhaust gasses does create some back pressure.

I have done extensive testing on exhaust systems and mufflers in the search for more power and discovered that Borla mufflers usually cost the least horsepower while offering a relatively quiet exhaust note. Most Borla mufflers utilize the absorption technique for reducing the exhaust note. All Borla

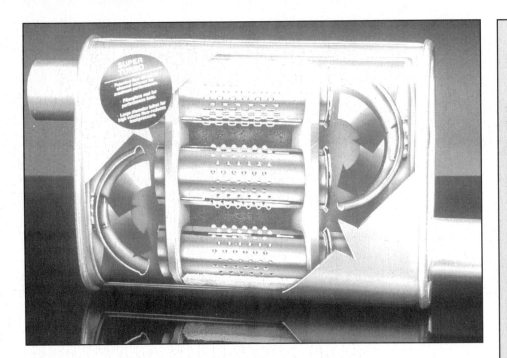

The Walker DynoMax series of performance mufflers also rate high in efficiency and at a reasonable cost. The Super Hemi Turbo muffler is both quiet and less restrictive. If you can fit them into your chassis the full offset, side in/side out mufflers are a little quieter than the center in/side out mufflers.

mufflers are made entirely of stainless steel which is also used as the absorptive material instead of fiberglass. The stainless steel is shredded into what looks like a large, loose Brillo pad that covers the pipes placed inside the muffler. Another unique Borla design on some mufflers includes using three or more smaller pipes inside the large inlet pipe. These smaller tubes not only increase exhaust velocity but also pipe area that can be exposed to the stainless steel matting, thus improving both exhaust flow and noise reduction. The only real downside to the Borla mufflers is their price. The lifetime warranty stainless steel does make the Borla some of the highest priced mufflers on the market. But if you're looking for ultimate power and durability, I believe these are among the best available.

If the stainless mufflers are a bit too pricey, there are a number of mufflers that are a bit more affordable that work well. As mentioned, the Flowmaster Pressure Busters are perhaps the next best buy. The Walker Super Turbo mufflers, especially the larger Hemi Super Turbo,

are an excellent compromise between cost and performance.

EXHAUST AND THE EPA

The advantage of building an older hot rod is that you don't have to worry about emission controls or restrictions on modifications. However, that's only on cars built before 1968, or 1966 in California. The cars with the most restrictions are cars built after 1975 that are equipped with catalytic converters. The Environmental Protection Agency (EPA) has produced guidelines that require manufacturers of aftermarket performance parts to do their own parts certification work to ensure that any component used on an emissions-controlled vehicle will not increase exhaust emissions from that vehicle.

California has an even stricter set of codes that requires manufacturers to apply to the California Air Resources Board (CARB) for an exemption, but only after the company has performed the necessary emissions certification work. If the component or combination of components pass this test, then the

PRESSURE TEST

One way to evaluate an exhaust system is to measure the amount of back pressure in the exhaust under wide-open-throttle conditions. An easy way to do this is to tap into the exhaust system just ahead of the muffler with a small pressure tap. Most exhaust pipes are made of sufficient wall thickness tubing that allows you to drill a 5/16-inch hole followed by an 1/8-inch pipe tap to cut threads in the pipe. Then use a standard brass fitting like the ones used for oil pressure fittings in the back of a small-block to secure a 1/8-inch copper tube. Using compression fittings for the connections, route this copper line up to a 0-15 psi fuel pressure gauge that can be mounted anywhere in the car. You must use copper line for this test. The plastic oil pressure lines will melt once the exhaust comes up to temperature.

With the gauge connected, the best test is done at wide open throttle since even stock exhaust systems rarely create much back pressure at part throttle. For safety, you should perform this testing on a track where you can run at wide open throttle long enough to obtain a reading. Maximum back pressure will occur at wide open throttle at peak horsepower so place the gauge on the dash or cowl where you can observe the gauge. Small amounts of back pressure will occur in any muffled system, so pressure readings of around 1 psi are actually low and don't represent a horsepower detriment. Back pressure readings between 4 and 5 psi are not outrageous. Back pressure above 5 psi indicates exhaust restriction that might be worth some power if it is reduced. Changing mufflers and/or improving the entire exhaust system will improve performance and fuel mileage.

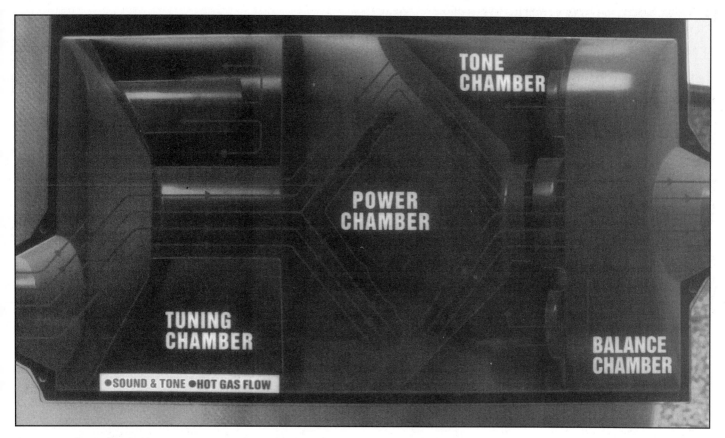

TONE CHAMBER

POWER CHAMBER

TUNING CHAMBER

BALANCE CHAMBER

● SOUND & TONE ● HOT GAS FLOW

Another muffler option is the new Pressure Buster mufflers from Flowmaster. These mufflers do not use absorptive materials, but rather wave cancellation techniques to dampen exhaust noise. This latest version of the Flowmaster is substantially quieter and worth more power than the original designs of ten years ago.

piece(s) can be legally sold for use on an emissions-controlled vehicle in California and are given an Executive Order (E.O.) number. Normally, any component recognized by the CARB will also be in compliance with 49-state EPA rules as well.

In regards to exhaust systems, both the EPA and CARB acknowledge that modifications performed downstream (after) of the catalytic converter are not subject to these rules. Therefore, high performance exhaust pipes, mufflers and tailpipes and kits like the popular "cat-back" systems do not require CARB E.O. numbers. This means that Borla, Walker or Flowmaster exhaust pieces after the cat are completely legal to use on any high performance emissions controlled vehicle.

In order to build performance packages for the late-model Corvette, Camaro and Firebirds, we have created a number of combinations that not only make

excellent power, but also do so without increasing emissions beyond the factory standards. In fact, in some cases the cars are cleaner at part throttle with the modifications than they were stock. For example, our shop has produced an emissions-legal LT1 package using the factory headers and cats that increases torque from 2000 rpm to 6000. For 6-speed cars, the stock intake is modified to help make 412 horsepower at 6000 with 385 lbs-ft. of torque at 4500. Most of this gain is achieved with the combination of cylinder head porting, camshaft and exhaust work.

If maximum torque is your goal, we can install the SuperRam for automatic transmission applications which boosts the torque to an astounding 425 lbs-ft. at 4000 rpm while in either case producing a smooth idle and excellent driveability. Obviously, these achievements didn't come easily, but they do prove that significant power gains are possible while

still achieving clean and legal exhaust.

EXHAUSTIVE CONCLUSIONS

It should be apparent now that matching the performance of the exhaust system to the rest of the engine is the best way to ensure the engine will make the most power it's capable of producing. Cutting corners on the exhaust side will only limit the amount of power the engine will make. For street-driven engines, a well designed high performance exhaust system is so efficient now that taking the time to "uncork" the exhaust is worth little in drag strip e.t. or speed. Oftentimes, a street car will run the same e.t. whether muffled or not. The existence of the Fastest Street car competitors running 7-second e.t.'s with mufflers merely underscore that statement. Pay attention to exhaust tuning and you'll go fast without having to be loud about it. ■

16 EFI AND SUPERCHARGING

Chevrolet switched to Electronic Fuel Injection (EFI) in the mid-'80s in order to more accurately control fuel into the engine for emissions reasons. Performance engine builders like myself quickly realized that this more precise fuel control could be used to make more power while improving driveability, emissions and fuel economy. While many of our engine packages are carbureted, this more accurate control of fuel and spark allows us to safely run the engines a bit leaner, which results in more power.

As we saw in Chapter 14, multi-point EFI allows more freedom in manifold design. This extends to supercharging as well. While supercharging has been around since the earliest days of the automobile, EFI allows more freedom in supercharger design, which has resulted in a growth in the number of different styles of superchargers as well as a surge in popularity of turbocharging. This growth is spurred by the ease with which an engine builder can now control fuel. While Roots blowers have been around since the '50s, multi-point EFI has opened the door for centrifugal superchargers and turbochargers for street engines. This can be seen in the growth of factory combinations of turbochargers and multi-point EFI. In this chapter we'll investigate the potential that awaits the ambitious engine builder willing to invest in the

"Positive manifold pressure," or forced induction, is the premise behind supercharging. This age-old power producing principle works well with today's EFI systems.

power possibilities of positive manifold pressure.

BLOWER BASICS

The original Roots supercharger used on GMC diesel trucks was not designed by GM. The Roots brothers designed the blower as a ventilation device for underground mines during the 1800's! The 6-71 designation was created by GMC for their diesel engines. The 6-71 designation describes a blower sized for a 6 cylinder diesel engine displacing 71

cubic inches per cylinder, or 426 cubic inches. The 4-71 blower case is smaller in overall size since it was intended for a four cylinder, 284 cid engine.

The Roots got its name "blower" because it was designed as an air mover, not an air compressor. The Roots blower creates pressure in an engine's intake manifold by increasing the rotor speeds to move more air into the intake than the engine can use. This results in air "stacking up" in the manifold, creating positive manifold pressure. But since the

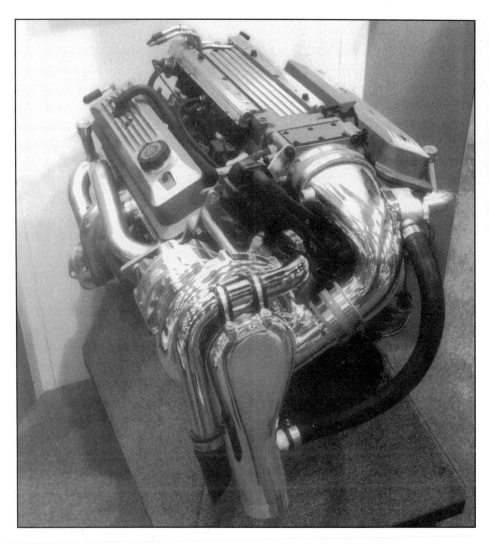

The combination of supercharging and EFI can create a powerful duo. This is a Vortech belt-driven centrifugal supercharger installed on a new LT1 engine. One advantage of the centrifugal supercharger is that it is easily installed under the stock hood on most fuel injected engines.

The Paxton centrifugal supercharger has been around since the mid-'50s. Recent modifications to the classic Paxton, identified as the SN-93, make it more durable than its predecessors. These units use an internal ball drive used to step up impeller speed.

blower was not designed as a high pressure pump, this manifold pressure finds its way back into the blower case where it is recirculated. This leakage reduces efficiency by creating additional heat. Despite its poor adiabatic efficiency, the Roots blower continues to be popular with the Pro Street set mainly because of its visual appeal and drag racing image.

In a normally aspirated engine, the pistons create a vacuum (a negative pressure) in the chamber. Atmospheric pressure then pushes air mixed with fuel into the chamber when the intake valve opens. The higher the atmospheric pressure, the more air will be shoved into the chamber. This is why all engines make more power at sea level than at 5000 feet of altitude. Taking this idea one step further, if you could use a mechanical pump to pressurize the intake manifold, it's possible to stuff more air and fuel into the chamber than with mere atmospheric pressure. The result is more power.

There are many different ways to create positive manifold pressure. The Roots supercharger is probably the most familiar to hot rodders, but there are many other devices that also create boost that are much more efficient. The reason the Roots supercharger is so popular today is because it is easy to combine with carburetors. The other types of superchargers include belt-driven centrifugal superchargers such as the Vortech or Paxton, the screw supercharger made popular by Whipple and, of course, exhaust-driven turbochargers. These alternative superchargers represent more efficient ways to increase manifold pressure with the least amount of temperature rise.

There have been books written about the various forms of supercharging, so this chapter must cover this complex subject very briefly. This chapter will deal mainly with the advantages of EFI and how it combines with belt-driven centrifugal superchargers because they are not only the most efficient, but also because they are easily adapted to an EFI small-block Chevy. While the classic Roots supercharger looks impressive sticking through the hood of a Pro Street car, the fact remains that the Roots blower presents far too many detriments to make it popular for a daily-driven machine. Even the mini-blowers such as

The classic Roots supercharger is easily the most identifiable of all superchargers. This is a cutaway of the B&M two-rotor mini-blower intended for street use. These smaller blowers work well on mild carbureted street engines when used at around 6 to 8 psi.

the B&M and Weiand superchargers find it difficult to squeeze under a stock hoodline.

The chief advantage of belt-driven centrifugal superchargers is that they can fit within even the cramped confines of a '94 Camaro engine compartment without the necessity of hood scoops. Bolting on the combination of a Paxton or Vortech with EFI under the hood of an older Chevy supercar is virtually a breeze. Add quiet operation and you have a very efficient combination that is almost transparent to the driver except for the amazing power. That's why the Mustang 5.0 guys like them so much. Let's take a look at why one type of blower is more efficient than another and how that efficiency contributes to more power.

How Blowers Work

Before we get into supercharger design, it's best to understand why blowers improve power and what happens when air is compressed. A supercharger improves upon atmospheric pressure's push by creating pressure in the intake manifold to shove the air and fuel into the cylinders. However, even with

turbochargers, this is not a free proposition. All engine-driven superchargers require power to turn. Even turbochargers generate exhaust back pressure that hurts power slightly. Furthermore, anytime air is compressed, air molecules are forced to squeeze together, which creates heat.

Adiabatic Efficiency—Under ideal conditions, there is a certain temperature rise that occurs whenever air is compressed to a given pressure above atmospheric. In theory, the least amount of temperature rise possible for that pressure increase is referred to as "100 percent adiabatic efficiency." Realistically, compressing air with a mechanical device like a supercharger adds frictional heat to this process. This lowers the adiabatic efficiency of the supercharging process. If you've ever touched the air tank on a large, shop air compressor when it's running then you know that compressing air generates heat.

The best way to measure the efficiency of compressing air is to evaluate the density of the air in the manifold. Density is a term used to describe the combination of pressure and temperature

of the air that is compressed. While compressing air increases its pressure and density, the added heat simultaneously reduces density. Ideally, a supercharger designer's goal is to create a given pressure (or boost) while adding the least amount of temperature. This creates the highest air density in the intake manifold.

Ratings—All superchargers can be rated in terms of their adiabatic efficiency. Roots blowers are less efficient since they create a great deal of heat under boost, which reduces the adiabatic efficiency to only around 50 percent. Next best on the list is the screw compressor such as the Whipple unit with a rating of around 65 to 70 percent. This blower is, in fact, a true compressor but is still not as efficient as belt-driven centrifugal superchargers that can generate adiabatic efficiencies of up to 75 percent. These are rough approximations since efficiency changes with boost, but these numbers do serve to position the different blowers.

Boost—Boost, or positive manifold pressure, is another commonly misunderstood factor related to superchargers. Common sense would suggest that more boost should equate with more power, and up to a point, this is true. However, high boost levels usually result from higher supercharger drive speeds. Most superchargers operate most efficiently at a certain speed. As the blower speed is increased to create more boost, the efficiency of the blower suffers because of the heat generated. At some point in the boost curve, the heat decreases the manifold density to the point where additional boost is offset by the temperature rise and no additional power is created. Further increases in boost will result in less power rather than more.

SUPERCHARGER DESIGN

Belt-driven centrifugal superchargers first became popular in the late '50s when used on the Ford Thunderbird but were

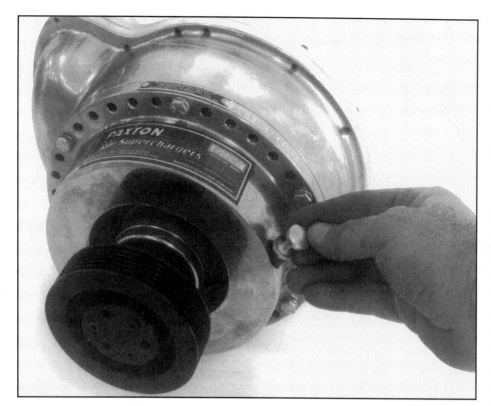

The Paxton uses an on-board oil supply to lubricate the drive mechanism with a dip stick for service. This makes installation easy but contributes to blower overheating under sustained use. Paxton has addressed this problem with a separate pump and cooler that keeps the oil cool but adds to the complexity of the system.

their way into the centrifugal supercharger market with dramatic results. This means that these overlooked superchargers are now coming into their own. The Vortech R- and S-trim superchargers are two of the leaders in this area.

Drive Mechanisms

One of the key items that differentiates the various centrifugal superchargers is their drive mechanisms. All centrifugal superchargers are driven off the engine crankshaft by a series of pulleys using a serpentine belt. All centrifugals also then use some type of internal set-up drive to further increase the impeller speed. The original McCulloch centrifugal supercharger used a step-up, ball drive mechanism still used in today's base Paxton blowers. The drive consists of five steel balls riding between two opposing races. One is driven by the shaft while the other is connected to a clutch pack that maintains pressure on the balls and is connected to the impeller. This ball drive increases the drive ratio of the input shaft rpm by a ratio of around 4.4:1. This

used in racing as far back as the 1920's. The original McCulloch design evolved into what is now the Paxton centrifugal supercharger. The centrifugal supercharger can be thought of as a belt-driven turbocharger since the design for compressing air is similar. The centrifugal supercharger uses very high speeds to spin an impeller that pulls air into its axis and then turns the air 90 degrees, flinging it at very high speeds into a scroll shaped much like a snail shell, called a *volute*. The increase in air speed created by the impeller creates a low pressure air that draws in additional air into the impeller, while the outlet air tends to "stack up" in the intake manifold since the centrifugal impeller is moving air faster than the engine can use it. The high impeller speed prevents back flow and therefore increases the adiabatic efficiency of the design. Less heat means more power.

Up until recently the design of the centrifugal blower's impeller blade was

impeded by both materials and design. However, recent improvements in turbocharger impeller designs have found

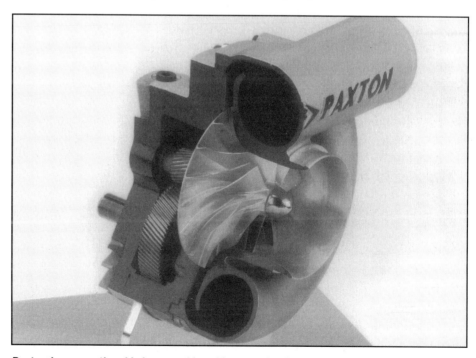

Paxton has recently added a gear-driven blower to its lineup called the Novi. Paxton claims increased boost levels with this supercharger compared to the ball-driven units. The Novi also incorporates a more aggressive impeller design that's responsible for the increased boost potential.

Impeller Design

In addition to the drive mechanisms, both Vortech and Paxton have refined their impeller designs to increase boost while also improving efficiency. Vortech has three different impeller designs. The original design is called the B-trim that is applicable for most applications making up to 450 horsepower. The next step up is the S-trim impeller that is capable of 20 percent more flow with capabilities of up to 540 horsepower. Vortech's most recent accomplishment is the R-trim impeller that, when matched with an engine capable of higher boost levels and airflow, can approach 900 horsepower! While the larger Vortech is somewhat lazy at low engine speeds, it is still capable of superior power throughout the rpm range when proper fuel and spark curves are maintained.

The newest Paxton is the Novi 2000 model that includes not only the improved helical gear like the V2 Vortech but also includes a highly refined impeller than is capable of serious power. This latest Paxton is new enough that we have not yet had the opportunity to test it. Most of our experience has been with the earlier Paxton SN-93 ball drive unit with

Vortech uses pressurized oil from the engine to lubricate the blower drive. This requires tapping the engine oil system which is easy, but also requires a return, usually through the valve cover.

design was quiet but tends to suffer from durability problems. Part of this stems from the Paxton's small-volume lubrication system. Paxton now offers an external pump and cooler system that increases the volume and also reduces the oil temperature.

Gears—The latest Paxton uses a helical cut gear drive on the Novi 2000 model. While this style of drive does increase the longitudinal force of the gearset, it is quieter than a straight spur-cut gear. The original V1 Vortech "Gearcharger" drive used spur or straight-cut gears to drive the impeller. The later V2 model now uses helical-cut gears that dramatically reduces gear drive noise. For lubrication, Vortech taps into the engine's oiling system to lubricate the drive gears. This is an advantage over the Paxton's dedicated oiling system. In both cases, hot oil contributes to increasing the operating temperature of the blower, adding to the temperature rise of the air leaving the supercharger.

The key to centrifugal superchargers is the design of the impeller. Early centrifugal blowers used straight impeller blades which were slow to build boost and not very efficient. Turbocharger technology is now being applied to the latest generations of centrifugal blowers. This is a B-trim Vortech.

its less efficient impeller.

Whipple Screw

A departure from both the centrifugal or Roots-style superchargers is the

Centrifugal superchargers rely on turning the impeller at very high speeds to generate boost. One way to increase the speed of the impeller is with a smaller drive pulley located on the snout of the blower.

Whipple screw supercharger. This blower is built by a company in Sweden and marketed exclusively in the United States by Whipple. The screw supercharger gets its name from the internal rotors that look like a pair of coarse thread screws. Air is pulled in from the back of the supercharger and compressed between the threads of the twin screws before being discharged from the bottom of the blower into the intake manifold. Since the screw supercharger is a true air compressor, it does not suffer from back flow problems and therefore is more efficient than the Roots supercharger. Cubic inch displacement is limited in these superchargers, which means they are best used in stock engines such as 305 and 350 cid applications. But like the Roots blower, the Whipple does offer immediate boost, making it more responsive than the centrifugal blowers.

EFI AND BLOWERS

Positive manifold pressure is perhaps the easiest way to generate both torque and horsepower. You can liken it to bolting on cubic inches since the additional volumetric efficiency is roughly like adding additional displacement. The biggest drawback to either turbochargers or centrifugal superchargers in the past was the carburetor. All centrifugal blowers and turbos work best when used in a "blow through" configuration where the blower pressurizes air and then pushes this air through the carburetor. The advantage is that the carburetor remained in its stock position over the engine.

Unfortunately, carburetors were never intended to operate in a positive pressure environment. The typical blow-through carbureted installation requires a sealed box around the carburetor that is difficult to work around since all connections have to be tightly sealed. Plus, fuel pressure has to be "boost referenced" to maintain that same 5 to 6 psi over the boost pressure. All of these problems limit the success of blow-through blower applications. With EFI, all of these limitations are instantly eliminated.

Installation

With EFI, bolting on a centrifugal supercharger quickly becomes the easiest way to increased power. The Vortech, Paxton and B&M kits merely bolt to the front of the engine and are belt-driven off the crankshaft. Then it's a simple job to plumb the inlet to a cold air source and

ACCEL/DFI FUEL MAP

LOAD (%)	.4			1.6			3.2			4.8			6.0			
0	18	18	18	18	20	20	27	28	30	31	32	32	31	31	30	30
	22	22	22	22	22	22	33	35	36	38	39	39	38	37	37	36
	25	25	25	25	25	25	39	41	43	45	46	46	45	44	43	43
25	30	30	30	30	30	30	45	47	50	52	53	53	52	51	50	49
	33	33	35	35	35	35	51	53	56	59	61	60	59	58	57	56
	35	35	40	40	53	56	59	62	65	68	70	69	68	67	66	65
	40	40	45	58	62	66	69	73	77	80	83	82	80	79	78	76
50	45	48	50	67	72	76	80	84	88	93	95	94	92	91	89	88
	62	67	71	76	81	86	90	95	100	105	108	106	104	103	101	99
	67	72	77	82	87	92	97	103	108	113	116	114	113	111	109	107
	72	78	83	89	94	100	105	111	116	122	126	124	122	120	118	116
75	77	83	89	95	101	107	113	119	125	131	135	133	131	128	126	124
	83	89	96	102	108	115	121	127	134	140	144	142	140	137	135	133
	88	95	102	108	115	122	129	136	142	149	154	151	146	144	140	136
100	93	101	108	115	122	129	137	144	151	158	163	160	155	150	145	140

ENGINE SPEED (RPM X 1000)

This is a fuel map used in an ACCEL stand-alone EFI computer. The vertical axis is load as expressed as a percentage of throttle opening with idle at the top and wide open throttle at the bottom. The horizontal axis is rpm with the far left low rpm and the far right maximum rpm.

HOT TIMES WITH HIGH PRESSURE

On another test, we recorded inlet versus outlet temperatures for the Vortech B-trim centrifugal supercharger. As you can see, at the lower boost levels, the centrifugal supercharger does not generate a serious temperature rise. However, as boost levels reach even a mild 7 psi, the blower outlet, or discharge temperature reaching the engine has doubled. Remember, manifold density is the key to more power. Excessive inlet air temperature lowers manifold density.

Manifold Pressure (psi)	Blower Inlet vs Outlet (Inlet/Outlet in degrees F)	Temperature Rise (Percentage)
2	75/94	25%
3	71/101	42%
5	70/120	71%
6.5	70/129	84%
7	71/140	97%
10	71/159	124%
11	71/174	145%

hook the blower outlet to the throttle body inlet of the EFI manifold. Since all the multi-point TPI, Lingenfelter SuperRam and new LT1 manifolds are "front breathers" with the throttle body placed toward the front of the engine, plumbing the centrifugal supercharger is relatively easy.

A positive pressure Manifold Absolute Pressure (MAP) sensor must be used along with an engine computer that is designed to map fuel with positive manifold pressures. Factory GM engine computers don't have this feature. A typical stand-alone computer application would use the ACCEL/DFI electronic control unit (ECU) to control both the spark and fuel requirements for the engine throughout both normally aspirated and boost conditions. We discussed the general characteristics of electronic fuel injection on page 120 in Chapter 14.

The combination of EFI and a centrifugal supercharger is by no means inexpensive. Each of these systems can cost as much as $3000 or more. But the advantage is that killer big-block power can be generated in a small-block that offers docile street manners. As you'll see in the accompanying sidebar, I built a 600 horsepower 383 that idles smooth with a power curve that makes 340 lbs-ft. of torque at 1600 rpm, cranks 589 lbs-ft. at 5000 rpm and over 600 horsepower at 6000 rpm. I have made this kind of power in larger, normally aspirated small-blocks, but these engines are usually rough-idling thumpers with long duration cams and high static compression ratios. The Vortech engine used a very mild Lingenfelter hydraulic roller cam and 8.8:1 compression yet still managed this amazing power.

As mentioned in the induction chapter, EFI does not, by itself, offer tremendous power gains over a well designed carbureted induction system. Manifold design is much more responsible for any significant power improvements. But when you add power by means of a centrifugal supercharger or a turbocharger, then the only game in town worth playing is EFI. As you can see, there are many interesting possibilities for the small-block when you combine high tech EFI and efficient supercharging. Either can contribute to a powerful small-block, but combining the two has proven to be extremely rewarding. ■

Both the Vortech and Paxton offer different outlet designs depending upon the installation requirements. The unit on the left requires the blower discharge to make a 90-degree turn at the outlet while the blower on the right discharges the air straight out of the blower. The unit on the right would be slightly more efficient as long as the downstream plumbing did not make any tight radius turns into the engine.

PRESSURE PLAY

Although Lingenfelter Performance Engineering is best known for our powerful normally aspirated small and big-block Chevys, we have experimented with a few supercharged small-blocks. This particular package was a 383 cid test mule used to create a docile small-block capable of awesome power. How much is awesome? Try over 600 horsepower at 6000 rpm, with 589 lbs-ft. of torque at 5000 rpm with only 11 psi of boost from a B-trim Vortech. The static compression ratio started at 8.8:1 while using a set of Lingenfelter-ported LT1 cylinder heads and the Lingenfelter/ACCEL SuperRam intake manifold. Cam timing was controlled by a Lingenfelter hydraulic roller camshaft developed specifically for supercharged engines. The test was performed using cast iron factory LT1 manifolds.

This combination has not been emissions-certified but it does offer tremendous opportunity for anyone interested in outstanding power and OEM driveability. In-car testing revealed that high under-hood temperatures prevented using 92 octane gas at the higher boost levels, with 7-8 psi the boost limit to avoid detonation. The high boost ceiling is limited by inlet air temperature and octane. With a higher octane fuel, more boost is possible.

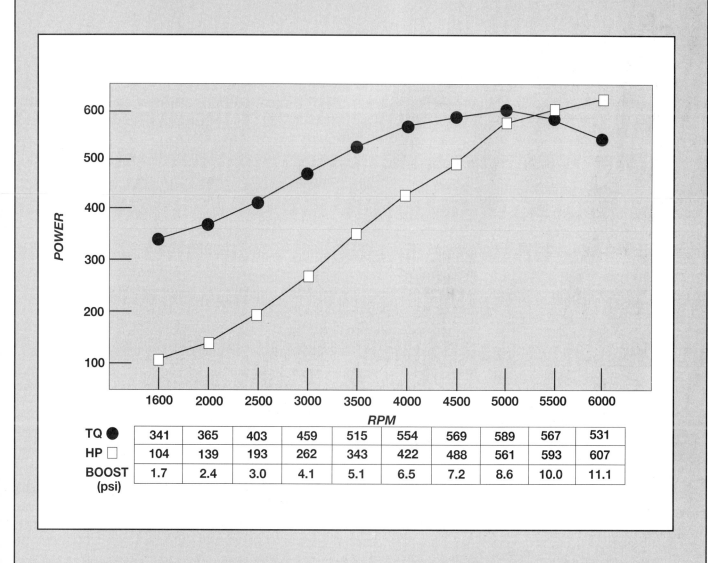

	1600	2000	2500	3000	3500	4000	4500	5000	5500	6000
TQ ●	341	365	403	459	515	554	569	589	567	531
HP □	104	139	193	262	343	422	488	561	593	607
BOOST (psi)	1.7	2.4	3.0	4.1	5.1	6.5	7.2	8.6	10.0	11.1

17 ENGINE BLUEPRINTING

One advantage that any beginning small-block Chevy engine builder has is that a few hundred thousand builders have gone before him, making his job much easier. Certainly no engine has ever had so much written about it, so there is plenty of material to help you when it comes to assembling your small-block. This chapter will outline some suggestions that will improve the longevity and durability of your small-block.

Many of the tasks outlined in this chapter have evolved out of careful attention to detail. These efforts also require specialty tools. If you're just starting out, many of these tools may seem out of reach for the casual engine builder. While purchasing these tools may seem an extravagance, that does not mean that you cannot get the job done. Find a machinist or enthusiast in your area that has these tools. Perhaps he can do the measurements for you and you can trade out the work for something you can do for him. Borrowing, renting or somehow making these connections are important since quality parts alone don't make a great engine. Double-checking all the clearances and ensuring that everything fits properly is the hallmark of a professional engine builder. Remember that the ultimate responsibility for an engine resides with the engine builder. Checking clearances yourself rather than

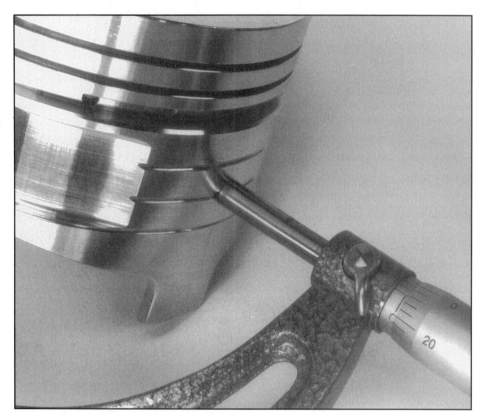

When it comes to high performance engine building, you can't afford to leave anything to chance. This includes checking all measurements and clearances yourself prior to assembling your engine.

relying on your machinist will surely lead to trouble at some point. If you check it before the engine is assembled for the last time, then you know it's right.

We'll look at a few of the more critical operations placed into two main categories—*Shortblock Blueprinting and Valvetrain Blueprinting*. While performing all these procedures is not mandatory for assembling an engine, they

establish a knowledge base so that when it's time to make changes to the engine, you have accurate information from which to start. This is especially helpful when disassembling the engine. Taking the extra time to measure as many clearances as possible will eliminate errors and result in a quality engine that will last.

When checking main or rod bearing bore diameters with a dial bore gauge, always measure the inside diameter (i.d.) with the dial bore gauge vertical or perpendicular to the load in the bearing bore. Bearings are built with a certain amount of taper as they approach the parting lines.

The only way to accurately measure cylinder bore for size, taper and out-of-roundness is with a dial bore gauge. The more accurate gauges will read to .0001 inch.

SHORTBLOCK BLUEPRINTING

Checking Clearances

The only proper way to measure bearing clearances is to install the bearings in the housing bore, torque the fasteners and measure the inside diameter with a dial bore gauge. Dial bore gauges are usually more accurate than an inside micrometer or snap gauges. Then measure the crank journals with the same micrometer used to set the dial bore gauge. This eliminates any variation between the micrometers. When torquing a connecting rod, make sure the rod is torqued in a rod vise. This prevents damage to the rod.

Don't Forget the Plug

Most shops remove this 1/2 inch oil passage plug under the rear thrust main cap to thoroughly clean the block. Be sure to replace the plug before the engine is assembled to prevent an internal oil leak. If the plug is not replaced, the engine will lose as much as 20 psi

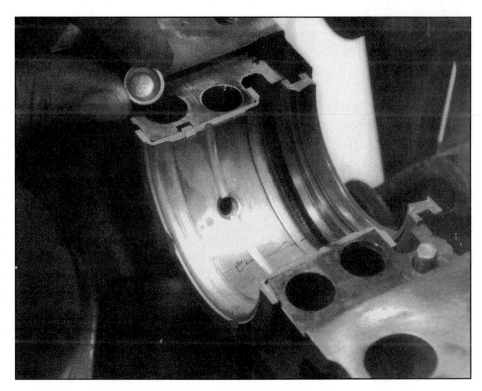

Don't forget this little 1/2-inch oil galley plug found under the rear main cap. If left out, it will create an internal oil leak that can reduce oil pressure by 20 psi.

because oil is leaking past this plug before reaching the rest of the engine. Idle oil pressure will be especially low. This missing plug can be mis-diagnosed

as a weak oil pump, excessive clearances or any number of problems that will defy repair.

Thrust Bearing Plate

High static pressure plate loads or high stall speed converters can really abuse a small-block's thrust bearing. One way to ensure it receives adequate lubrication is to place a small, .020 inch chamfer on one rear side of the upper thrust bearing half. This chamfer channels a small amount of oil from the oil groove in the upper bearing half to the rear thrust bearing surface that could prevent thrust bearing problems. Also paint the mating surfaces of both the main cap and block with a light coating Permatex sealant to prevent oil from leaking past the main cap.

Line-Up—Another way to ensure long thrust bearing life is to take a little extra time when installing the #5 main bearing cap for the last time. Place the main cap and thrust bearing on the block and lightly tighten the main cap bolts. With a clean pry bar, push the crank rearwards. Now pry the crank forward and then torque the main cap bolts to their required torque spec. Prying rearward and then forward aligns the thrust bearing surfaces. Prying forward as the last step ensures the rear thrust surfaces are aligned for a complete bearing surface since any high thrust forces will come from the rear pushing forward on the crank.

Rod Side Clearance

While some books show you how to measure rod side clearance by measuring the thickness of the paired rods and the width of the crank journal, this cannot be done accurately. Production rods are often not completely flat, which will create a tight spot in one particular spot. The only way to measure rod side clearance is to mock the rods up on the crank with a set of checking bearings and measure the side clearance with a feeler gauge. Be sure to measure all the way around the rod circumference to ensure there are no tight spots.

Chamfering this edge on the thrust bearing will supply a small amount of pressurized oil to the thrust bearing.

End play is easily verified with a dial indicator placed on the snout of the crank. Pry the crank forward and back and check the clearance.

Aligning the rear thrust main bearings will not only improve thrust bearing life but will also affect crank endplay. Check endplay only after the rear thrust bearing has been aligned.

Make sure you check rod side clearance completely around the entire circumference of the rods. Sometimes high spots on a rod will create a tight area that could create heat.

Deck height is the relationship of the piston to the deck surface of the block. This is easily checked with a dial indicator and a deck bridge.

Ring End Gap

Measuring this is as simple as placing the rings about 1/2 inch down in the bore and squaring them with a flat top piston. Then measure the clearance with a feeler gauge. Most professional engine builders prefer to use .005-inch oversize rings so they can file-fit the rings to the end gap they prefer. When using a ring filer, work slowly and deliberately. Sneak up on the desired clearance since it's very easy to remove too much. Once the ring end gap is set, be sure to deburr the filed end with a light whetstone. Always deburr by filing away from the end of the ring rather than into it.

Deck Height

This is a relatively easy procedure to check, but it involves mocking up all eight pistons and rods since each must be checked. You do not need to install the rings. Once all eight pistons are in place, the easiest way to check deck height is rotate the engine over with a bridge type dial indicator positioned over the cylinder. Place the bridge so that the dial indicator will reference off the piston on the flat surface, not off a dome or dish. Also make sure the indicator is located over the wrist pin. This diminishes the effect of piston rock in the bore at TDC, which can be as much as .002 to .003 inch.

Piston-to-Valve

Even flat top pistons with big cams can run into valve-to-piston clearance problems. The easiest way to check this is to mock up one cylinder with the crank, bearings, rod, piston, cam, valvetrain and cylinder head. Place a small amount of clay in the piston valve reliefs and then install the cylinder head. If you don't have a similar used head gasket, you can leave the gasket out and then add the gasket thickness into the measured clearance.

Bolt the head with the five bolts around the cylinder and lightly tighten them in place. Assemble the lifters, pushrods, rocker arms and valves. If your engine will be equipped with hydraulic lifters, you must substitute solid lifters for this test since hydraulics will bleed down and produce a false reading. Set zero lash for both the intake and exhaust and then gently turn the engine over twice. This will put both the intake and exhaust valves through their entire cycle. If the engine stops abruptly, don'f force it! The valves may be hitting the pistons. Remove the head and use a machinist scale to measure the thickness of the clay. This is the piston-to-valve clearance. Don't forget to add the head gasket thickness if you didn't use one on the engine.

Oil Pump Pickup Height

Most pump manufacturers recommend that the oil pump pickup be placed between 3/8 and 1/2 inch from the bottom of the oil pan. If the pickup is too far from the bottom of the pan, this could uncover the pickup during hard acceleration or high lateral g loads. This also reduces the effective volume of oil in the pan. A pickup placed too close to the bottom of the pan restricts the pickup and could cause oil pressure problems at high rpm.

The easiest way to check this clearance

ENGINE CLEARANCES

The following are engine clearance recommendations based on my experience in building street and race small-block Chevys. One chart cannot begin to list all the different clearances for all applications. This chart will assume the application to be a daily-driven street small-block that will run on 92 octane pump gasoline with steel connecting rods and forged pistons. Consider these clearances to be general recommendations rather than hard numbers. If you have questions, refer to the particular chapter or call the manufacturer of the parts in question.

CLEARANCE SPECS

COMPONENT	CLEARANCE (inches)
Main Bearing:	.002-.0025 Mains 1-4, #5 .0025-.0030
Rod Bearing:	.002-.0025
Rod Side:	.009-.013
Crankshaft Endplay:	.003-.010
Piston-to-Wall:	Refer to Manufacturer
Ring End Gap:	Refer to Manufacturer
Ring Side Gap:	.001 Refer to Manufacturer
Piston-to-Head:	.040 (steel rod)
Valve-to-Piston:	I = .080 E = .100
Valve-to-Guide:	I = .0015 E = .0015 - .0020
Retainer-to-Seal:	.050
Coil Bind:	.050

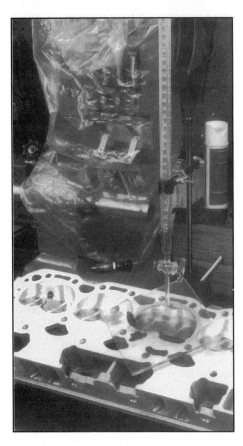

I use this 100cc burette to measure chamber volumes. The stand keeps the burette steady and makes measurement easier. The fluid is usually rubbing alcohol mixed with a food dye. This makes cleanup easier.

is to place a small ball of clay on the pickup with the engine upside down on an engine stand. Then place the oil pan rail gaskets on the engine, and install the pan and bolt it down with a couple of bolts. Now remove the pan and measure the thickness of the clay. This will be the distance of the pickup from the bottom of the pan.

VALVETRAIN BLUEPRINTING

Computing Compression Ratio

All hot rodders know that the harder you squeeze the air and fuel in an engine, the more power you will make. There are limitations to this of course, with the most important limitation being the octane of the fuel. For street engines, the benchmark has been the 9:1 rule. But like all generalizations, there are exceptions. I push this limit. I place compression at 9.2:1 on carbureted iron head engines and 10.2:1 on aluminum head carbureted engines. Our EFI engines stretch this even further. Regardless of the style of engine, the only way to know is to measure the volumes and compute the compression ratio.

Simply stated, the compression ratio is the volume of the cylinder with the piston at the bottom of its stroke compared to the volume of the cylinder with the piston at the top of its stroke. Unfortunately, we're dealing with cylinders that are not perfectly true. Crevice volumes around the rings, piston domes, valve reliefs, deck heights, head gasket volumes and combustion chambers all contribute to this formula.

Since we are working with a number of variables, the first thing we must do is ensure that all of our values are expressed the same way. Most hot rodders express displacement in cubic inches (ci's), but combustion chamber volume is generally given in cubic centimeters (cc's). To make the formula easier to work, we'll convert cc's to cubic inches. The conversion from cc's to ci's is:

$$CC \times .061 = CI$$

For example, a 76cc chamber is equal to 76 x .061 = 4.636 ci. Converting cubic inches to cubic centimeters is CI x 16.39

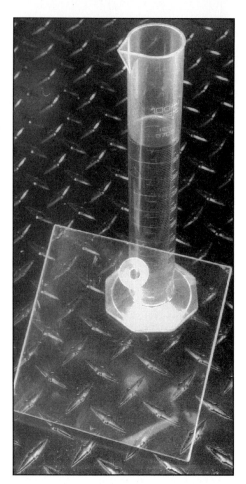

CDI now offers an inexpensive CC'ing kit for under $25. It's not as nice as the glass burettes, but far less expensive. You can use this tool for checking combustion chamber volume, and crevice volume to compute compression ratio, not to mention port volume and many other uses too.

= CC's. So 4.636 ci x 16.394 = 76 cc's.

With all these irregular volumes, the easiest way to determine the compression ratio is to compute each volume separately and then merely divide the larger by the smaller volume. The variables are: 1) overall swept volume, 2) combustion chamber volume, 3) compressed head gasket volume, 4) deck height volume and 5) piston volume. Piston volume can be either a value for a dish or a dome. For use in our formula, a dished piston will be a positive value greater than zero while a domed piston will be expressed as a negative value, subtracted from the others. This is because a domed piston reduces the overall volume of the combustion chamber while the dished piston adds to the chamber volume.

1. Let's start by computing the overall swept volume of the cylinder. This is a simple formula:

Bore x Bore x Stroke x .7854 = Swept Volume in cubic inches.

For example:

4.030 x 4.030 x 3.48 x .7854 = 44.389 ci. The constant .7854 is a shortcut that keeps the number of values down to a minimum.

2. The next value is combustion chamber volume. This is determined by measuring the volume of the chamber with a burette. But we must convert this value from cc's to ci's: CC's x .061 = Chamber Volume in cid's. For example, 76cc x .061 = 4.636 cid.

3. The thickness of the head gasket affects compression as well. If you do not have the compressed thickness volume, you could compute it by using the same formula used to compute swept volume. For example, a.038-inch thick head gasket for a 350 Chevy would compute:

4.03 x 4.03 x .038 x .7854 = .485 cubic inches.

Unfortunately, this doesn't work very well because the gasket companies usually build one head gasket to fit a variety of bore sizes. This means the cylinder bore opening in the gasket is far larger than 4.030. Fel-Pro's 1003 gasket for a 350 has a cylinder bore opening of 4.166 inches. Using this larger bore size computes to .518 cubic inches compared to .485 cid. Converting .518 cid to cc's equals 8.09 cc's. Fel-Pro's published volume is 8.7 cc's for this gasket which equates to .531 ci. This is because the bore opening is an irregular shape for valve reliefs, increasing the volume of the gasket opening.

4. Deck height is the distance the flat portion of the piston is below the top of the cylinder block, which affects compression just like gasket thickness. Most stock small-block Chevys measure between .020 and .025 inch below the top of the block. This volume is computed just like gasket volume. For 350 with a .025-inch deck height. The formula is:

4.03 x 4.03 x .025 x .7854 = .319 cubic inches.

5. An accurate compression ratio requires accurate measurement. Very few small-block pistons are perfectly flat. Most have valve reliefs or a dish that increase volume or domes that decrease volume in the chamber space. While many piston companies specify dome or dish volume in cc's, this doesn't take into account the area between the top piston ring and the top of the piston called the crevice volume. The best way to determine the total volume of the piston is to measure it in the cylinder.

Mock up one piston with a top ring in place. Smear a light coating of grease on the cylinder wall to help seal the ring and place the piston a measured distance down in the cylinder, like .250 inch. If you are using a flat top piston, you could reduce this distance to .100 inch to reduce the volume for ease of measurement. Use a deck height micrometer to establish this distance accurately. Then, using the same clear plastic cover used for cc'ing combustion chambers for a lid, cover the cylinder and measure the amount of liquid required to fill this volume. The difference between the calculated volume of a perfect cylinder and the measured volume will be the volume displaced by the piston dish/dome/valve reliefs.

For example, computing the volume of a true cylinder of 4.030 inches with the piston .100 inch below the deck surface would be computed:

4.030 x 4.030 x .100 x .7854 = 1.275 cubic inches (or 20.90 cc's)

Measuring this same volume, we come

Competition Cams sells a slick camshaft degreeing kit in this handy plastic case that gives you everything you need to degree a cam. They also include a video and instruction sheet to make it easy to do it right the first time.

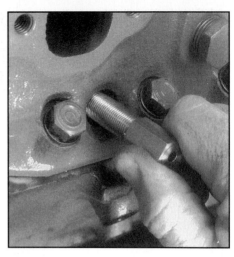

The first step is finding TDC. Bolt on the degree wheel and put the number one piston at TDC. True TDC is located when the piston stop numbers on either side of TDC read the same.

up with 25 cc's that converts to 1.525 ci. Subtracting the measured volume from the computed volume produces: 1.525 - 1.275 = .250 ci. This is the total volume of both the valve reliefs and crevice volumes.

Domed Piston—This is for a flat-top piston. In the case of a domed piston, the measured volume would be less than the computed volume. For example, let's say we use a domed piston in this same cylinder with a manufacturer's volume of 12 cc's. This converts to .732 ci. The actual piston volume would then be 1.275 - .732 = .543ci or 8.9cc's. Note that this volume must be subtracted from the combustion chamber volume in order for the formula to work properly since the dome becomes a negative volume.

Now that we've determined all the separate volumes, let's put them together into a formula that will determine the compression ratio. The V stands for volume.

$$\frac{\text{Swept. V.} + \text{Chamber V.} + \text{Gasket V.} + \text{Deck V.} + \text{Piston V.}}{\text{Chamber V.} + \text{Gasket V.} + \text{Deck V.} + \text{Piston V.}}$$

This example is a 350 with a 4.030 bore, 3.48 stroke, 76cc (4.636 ci's)

chamber, .038-inch (8.7 cc's = .531 ci) Fel-Pro gasket, .025-inch deck (.319 ci) and a measured piston crevice volume of .250 ci's.

$$\frac{44.389 + 4.636 + .531 + .319 + .250}{4.636 + .531 + .319 + .250} = \text{C.R.}$$

$$\frac{50.125}{5.736} = 8.75{:}1 \text{ compression ratio}$$

As mentioned in the piston chapter, reducing the deck height will improve combustion efficiency not just by increasing compression but also by enhancing turbulence in the chamber. Let's see what decreasing the deck height to .005 inch (.064ci) on our flat top piston example will do to the compression ratio.

$$\frac{44.389 + 4.636 + .531 + .064 + .250}{4.636 + .531 + .064 + .250} = \text{C.R.}$$

$$\frac{49.87}{5.481} = 9.1{:}1 \text{ compression ratio.}$$

Let's substitute a 12cc domed piston for the flat-top and see what happens to the compression ratio. Note that we are now adding a negative piston dome volume to

keep the rest of the formula the same.

$$\frac{44.389 + 4.636 + .531 + .064 + (-.543)}{4.636 + .531 + .064 + (-.543)} = \text{C.R.}$$

$$\frac{49.077}{4.688} = 10.47{:}1 \text{ compression ratio}$$

How to Degree a Camshaft

There is more to installing a camshaft than just slipping the cam in and lining up the marks on the timing gears. To degree a cam, you will need some basic tools to accomplish the task. Competition Cams, like other cam companies, offers a camshaft degreeing tool that includes all the items you'll need to complete the degreeing process. Of course, you can also assemble your own set of tools that should include a degree wheel, dial indicator and magnetic base, piston stop and a coat hanger wire pointer.

The first task in the degreeing process is the most important—establishing Top Dead Center. The first thing to do is place Number One piston at Top Dead Center and then install the degree wheel on the crankshaft. You'll also need some type of crankshaft nut, sold by various aftermarket companies like B&B, to turn

The Comp Cams kit designs the system to mount the dial indicator on the valvespring retainer. Rotate the engine through a number of cycles to ensure the dial indicator always returns to zero when the lifter reaches the base circle.

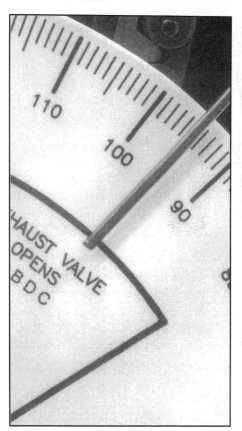

When degreeing a cam by the intake centerline method, first record the cam timing figure for the intake lobe at .050 inch before maximum lobe lift. In our example, we have 95 degrees ATDC.

Then turn the engine until you record .050 inch after maximum lobe lift. Once you have these two numbers, add them together and divide by 2 and you will have the intake lobe separation angle. In this case, the second .050 number is 115 degrees ATDC. The result is 105 degrees ATDC. The spec for this Comp Cam is 106 ATDC, so we would retard the cam one degree to be dead-on.

the crank so as not to disturb the crank bolt securing the degree wheel.

With the degree wheel in place, position the degree wheel at TDC. To find the degree wheel's true TDC location, we will need a positive piston stop. If the heads are off, a piston stop plate is used over the bore using a bolt placed to stop the piston before it reaches TDC. If the heads are already installed on the engine, then a piston stop screwed into the spark plug hole will work just as well. These stops can be homemade or purchased from the aftermarket.

With the piston stop in place, turn the engine clockwise until the piston positively contacts the piston stop. Record the number on the degree wheel. Then turn the engine counterclockwise until the piston comes up against the stop.

Be careful not to damage the piston. Record the number on the degree wheel. When the degree wheel reads the same number of degrees on either side of TDC, then the degree wheel is precisely located at TDC.

For example, if the degree wheel reads 28 degrees before TDC (BTDC) and 20 degrees after TDC (ATDC), then either the degree wheel or the pointer need to be moved in order for TDC to be precisely located. To make this easy, add the two numbers together and divide by two to determine what the pointer should read on both sides of TDC. In this case, 28 + 20 = 48/2 = 24. The wheel needs to be moved four degrees so that the wheel would read 24 degrees on both sides of TDC. This is a critical step because all subsequent cam degree readings are

based on the accurate location of TDC. Always double-check the degree wheel's TDC to be sure it is located accurately.

With TDC located, place a lifter in the second lobe back from the front on the driver's side, which is the intake lobe. Position the dial indicator with its magnetic base so that the pointer will be in line with the lifter. Rotate the engine through two or three complete rotations to ensure that the dial indicator always returns to zero on the base circle of the camshaft. If the dial indicator does not return to zero, reposition the dial indicator until it repeats.

There are several ways to measure cam timing. Competition Cams uses the intake centerline method. Other methods include checking the opening and closing points at .006-inch or .050 inch tappet lift. We'll outline the intake centerline method since this involves a couple extra steps.

In the intake centerline method, rotate the engine clockwise until the dial indicator reaches maximum lift. For a mild street cam, let's say this figure is .300 inch. Move the dial on the indicator until it reads 0. Now rotate the engine backwards, or counterclockwise until the dial indicator reads .100 inch or so less than maximum lobe lift. Now rotate the engine clockwise again until the degree wheel reads .050 inch before 0 or maximum lift. Record the reading on the degree wheel. For this example, let's say this is 95 degrees After Top Dead Center (ATDC).

Now rotate the engine in the same clockwise direction until the dial indicator reads .050 inch past maximum lobe lift on the closing side of the lobe. In our example, this number is 115 degrees ATDC. Add the two numbers together and divide by two: 95 + 115 = 210 / 2 = 105 degrees ATDC. This is the intake centerline of the intake lobe. This is within one degree of the 106 degree ATDC intake centerline figure for the camshaft. To obtain 106, we would retard the cam one degree.

If the cam measured at 109 degrees ATDC, then it is considered to be late, or retarded. If the cam measured 104 degrees ATDC, then it would be early, or advanced. Advanced cam timing means both intake and exhaust valves open earlier than specified by the cam grinder. Conversely, a retarded cam opens the valves later. Advancing a cam usually results in some slight gains in low and mid-range torque at the sacrifice of top end power while the opposite is true of a retarded cam.

Moving the cam is accomplished by a number of different methods. There are small eccentric bushings that fit between the cam's locating pin and the cam gear. These bushings come in 1, 2, 4 and 6-degree increments to allow the engine builder to custom position the cam timing. Also popular are the three- and sometimes four-way crank gears that come with multiple keyways marked for advancing or retarding the cam timing. Many of the timing chain set manufacturers like Cloyes now offer adjustable timing gears that can change cam timing by merely loosening an Allen set screw and moving a timing mark to the position you require. These are handy, but also more expensive than production type timing sets.

This may sound like a complex operation, but in reality it goes fairly quickly. Once you've done it a couple of times, it becomes a simple task that you can breeze through without difficulty.

Setting Pushrod Length

There are many different ways to set pushrod length, but one quick way is to mock up the head with the valves at their established height and the springs in place. Lightly coat the top of the intake and exhaust valve with machinist's blue dye. Ensure that the cam lobe is on the base circle. Now install the pushrod and rocker arm. With a slight amount of pressure on the rocker, rock it across the face of the valve to establish a witness mark on the valve tip.

This mark should be on the inboard 1/3 of the valve stem tip if the pushrod length is correct. If the mark is in the center of the valve stem, then the pushrod is too short. If the witness mark is farther inboard than roughly 1/3, then the pushrod is too long. You can experiment with this by using an adjustable pushrod from Competition Cams to come up with

Pushrod length is correct when the witness mark on the valve tip is approximately 1/3rd of the way in from the inboard side with the lifter on the base circle of the cam. Total travel of the rocker arm should be across the middle third of the valve stem tip, anywhere from .030 to .050 inch in width.

the proper pushrod length. To check this further, you can install a solid lifter, set the rocker to zero lash and run the valve through its entire lift cycle. Now look at the witness mark on the valve stem tip. It should reveal a wear pattern from the inside 1/3rd to the outside 1/3rd of valve tip.

The proper length pushrod will minimize this total roller tip travel across the face of the valve stem tip to approximately .040 to .050 inch or less. A pushrod slightly too long or too short will tend to increase this total travel distance. Proper pushrod length not only limits this frictional wear across the valve stem tip, but also decreases the side load imparted onto the valve stem, which improves valve guide durability for an endurance or street engine. ■

ENGINE DYNO TESTS 18

All the previous chapters have led up to this one. This is where the fun starts, where all the hard work of creating combinations of heads, cams, compression, intake and exhaust all come together. It's okay if you took the shortcut and came right to this chapter first. But keep in mind that reading all the previous chapters will make the next few engine combinations more clear and you'll understand why the parts were combined as they are.

In case you've yet to notice, I love the feel of torque. Torque is what accelerates the car. For street cars, torque is especially important at the lower engine speeds since this kind of power makes the car especially fun to drive even at part throttle. Peak horsepower is important, but with street cars increasing torque even at the sacrifice of some peak horsepower will make the car both more fun to drive and quicker in acceleration. The goal of this book is to illustrate how and why these combinations were created. If you learn these fundamentals, then building future engines will be easier, more fun and less confusing.

There is probably no limit to the variety of small-block Chevy street engine combinations. Personal preference, budgets and a dozen other factors will dictate these combinations. Out of all these mind-numbing options, I have developed several popular combinations.

This 400 cubic inch aluminum Bow Tie block with 18 degree Bow Tie heads and a killer roller cam is about ready to go to the dyno for extensive testing.

We'll split these combinations into three groups. The first group is carbureted engines in 355, 383 and 406 cid displacements. The second group will be multi-point, fuel-injected engines. The third group will be emissions-legal EFI combs for the late-model Corvette, Camaro and Firebirds.

The carbureted versions are less expensive than the EFI engine packages and usually make less torque depending upon the intake manifold. Note how displacement has a big effect on torque. You can increase torque by lengthening the intake runner and shortening the cam timing, but these tricks also limit horsepower. Displacement is one place

where bigger is better as long as fuel mileage is not critical.

The mild small-blocks typically use production style heads. Using the ported Corvette head or other aftermarket head on a mildly-cammed engine will generally increase both torque and horsepower. The combinations are limited only by your imagination. You may also note that some of these engines make more than the claimed power. I typically underrate many of my engines to ensure each engine always meets the claimed power levels. Now, let's fire up these engines, crank up the dyno and watch that beam indicator swing around to max power!

CARBURETED SMALL-BLOCK COMBINATIONS

355 cid
360 HP/375 LBS-FT.

    This engine package is intended for the street enthusiast who intends to drive the car on a regular basis. The combination produces a noticeably rough idle and could benefit from a slightly higher stall speed converter in automatic applications. I also offer a milder version of this engine making 310 horsepower and 360 lbs-ft. of torque with a smoother idle.

    The engine starts with a two-bolt main block, cast crank and rebuilt 5.7-inch rods. Sealed Power forged, flat-top pistons produce 9.0:1 compression while a single moly ring package, Clevite 77 bearings and a standard volume, high pressure pump complete the bottom end. A Competition Cams single pattern Magnum 280 hydraulic cam with 230 degrees of duration, .480-inch lift and a 110 degree lobe separation angle, a roller timing chain, hardened pushrods and stamped steel 1.6:1 rockers complete the valvetrain. To keep the cost down, this combination uses a set of Lingenfelter-modified 492 iron heads fitted with 2.02/1.60-inch valves with hardened seats. An Edelbrock Performer intake, chrome valve covers and timing chain cover and a stock pan complete the package. For the dyno test, this engine was fitted with a 750 double pumper Holley carburetor and a set of 1 5/8-inch headers. Peak torque occurs at 4500 rpm with peak power at 5500.

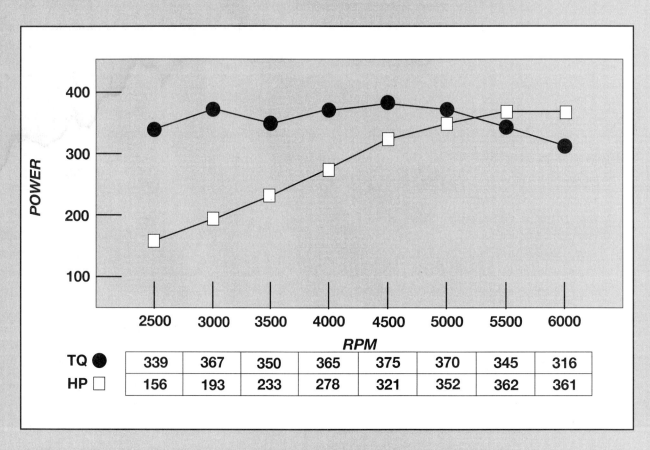

	2500	3000	3500	4000	4500	5000	5500	6000
TQ ●	339	367	350	365	375	370	345	316
HP □	156	193	233	278	321	352	362	361

355 cid
400 HP/ 390 LBS-FT.

This 355 is intended for the serious small-block enthusiast where a choppy, rough idle is acceptable. This combination will work best in a vehicle with a stall speed converter of at least 3000 rpm and deep gears to take advantage of the higher rpm power curve.

The foundation for this combination is a four-bolt main block, Chevrolet steel crankshaft and 5.7-inch Chevrolet rods. Sealed Power pistons produce 10.5:1 compression mated with single moly rings and Clevite bearings. The emphasis on horsepower moves cam timing up to a Competition Cams 292 Magnum hydraulic with 244 degrees of duration, .501-inch lift and a 110 degree lobe separation angle with matching lifters and a roller chain. Airflow is handled with a set of ported late model Corvette heads with 2.00/1.56-inch valves, guideplates, stiffer springs and 1.6:1 stamped rockers. The induction is a Victor Jr. intake along with a 750 Holley double pumper and 1 3/4-inch headers for exhaust. Note that even with this long duration camshaft that the engine produces a broad power band with torque in excess of 340 lbs-ft. from 3500 to 6250.

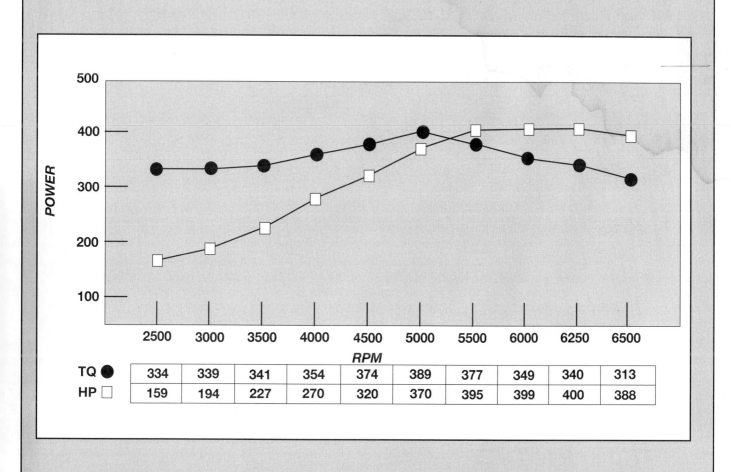

	2500	3000	3500	4000	4500	5000	5500	6000	6250	6500
TQ ●	334	339	341	354	374	389	377	349	340	313
HP □	159	194	227	270	320	370	395	399	400	388

383 cid
385 HP/410 LBS-FT.

This is the first step up to more displacement. We also offer a slightly less powerful 383 combination with a milder cam and iron heads that makes 325 horsepower and 395 lbs-ft. of torque with a smooth idle that makes a great street engine.

The 385-horse combo uses a two-bolt main 350 block, cast Chevy 400 crank and the matching 5.565-inch production rods. Sealed Power forged pistons generate 9.2:1 compression with moly rings, Clevite bearings and a standard volume, high pressure pump. A Competition Cams 280 hydraulic supplies the lift and duration along with a roller timing chain, hardened pushrods, guideplates and stamped steel 1.6 rocker arms. Heads are modified production iron castings fitted with 2.02/1.60-inch valves. This package is outfitted with an Edelbrock Performer RPM intake and was tested with a 750 Holley carburetor and 1 5/8-inch headers. The idle quality is slightly rough with the 280 cam, but still very streetable. This is a great street-torque combination that will run on 92 octane premium with no problem.

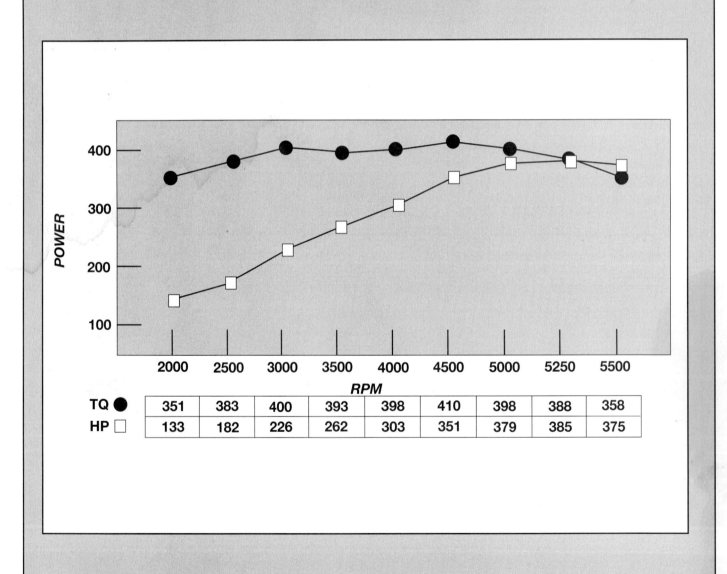

		2000	2500	3000	3500	4000	4500	5000	5250	5500
TQ ●		351	383	400	393	398	410	398	388	358
HP ☐		133	182	226	262	303	351	379	385	375

383 cid
475 HP/450 LBS-FT.

Now it's time to get serious. This is a thumper 383 with more power and torque. This combination will require a high stall speed converter of at least 3000 rpm, along with 1 3/4-inch headers, an excellent exhaust system, 92 octane pump gas and a Holley 750 carburetor. You'll note that while I state 450 lbs-ft. of torque and 475 horsepower for most engines, this particular combination made an astounding 478 lbs-ft. and almost 490 horsepower!

This 383 starts with a four-bolt main 350 block and a cast 400 Chevy crank but adds longer 5.7-inch 350 rods for an improved rod-length-to-stroke-ratio. Sealed Power forged pistons pump the compression to 10.5:1 along with moly rings, Clevite bearings and a standard volume, high pressure pump. We add a solid street roller Comp Cams grind with 236 degrees at .050-inch tappet lift with .587-inch lift to this package along with a roller timing chain and a new harmonic balancer. This also means a step up to the ported Corvette heads with 2.00/1.56-inch stainless valves, hardened pushrods and 1.6:1 rocker arms. A Victor Jr. and a Holley 750 double pumper provide the induction.

The solid lifter street roller cam will require periodic maintenance but the combination of these high velocity heads, the roller cam and the 10.5 compression make for a powerful package. This is an easy 12-second combination in a 3500 pound car.

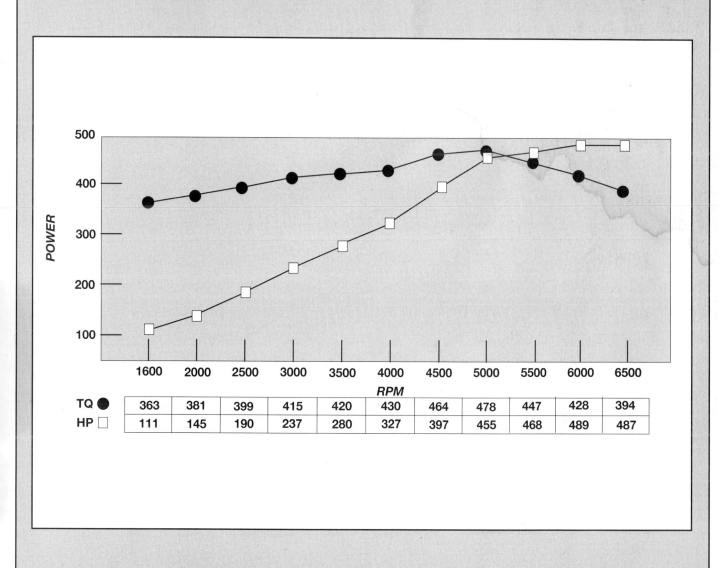

	1600	2000	2500	3000	3500	4000	4500	5000	5500	6000	6500
TQ ●	363	381	399	415	420	430	464	478	447	428	394
HP □	111	145	190	237	280	327	397	455	468	489	487

LINGENFELTER EFI SMALL-BLOCK COMBOS

The chapter on induction systems details the difference between the EFI style intakes and the typical carbureted induction systems. We designed the SuperRam to maximize the torque of a street-driven small-block, which this manifold delivers handsomely. Merely switching from a Performer RPM intake to a SuperRam with a properly tuned fuel and spark curve can be worth as much as 50 lbs-ft. of torque in the middle of the power curve. That is an impressive torque gain. Suspension companies love this kind of engine because after bolting this rascal in your biggest hurtle is how to control tire spin.

383 cid EFI
440 HP/480 LBS-FT.

This would make a fantastic street engine. There is also a milder version of this engine using the SuperRam but with a milder cam that will idle at 750 rpm yet still make 395 horsepower and 450 lbs-ft. of torque. Actually, calling this combination "mild" is a misnomer.

The short block for this 383 is similar to the 385-horsepower carbureted combination. This 383 uses a .030-over 350 block, cast 400 crank and 5.565-inch 400 rods, Sealed-Power forged pistons, moly rings and Clevite bearings. A roller timing chain drives a Lingenfelter hydraulic roller camshaft with 219 degrees of duration at .050-inch tappet lift, .503/.525-inch valve lift and a 112 degree lobe separation angle.

We then port a set of aluminum 'Vette heads with 2.00/1.56-inch valves adding matching valve springs and Comp Cams stainless steel 1.6:1 rocker arms. With the SuperRam and a 58mm throttle body on the intake side and 1 3/4-inch headers, this engine is a definite torque monster making 363 lbs-ft. of torque at 1600 rpm! Peak horsepower occurs at 5250 rpm. Like many of the other combinations, a loose converter would help acceleration.

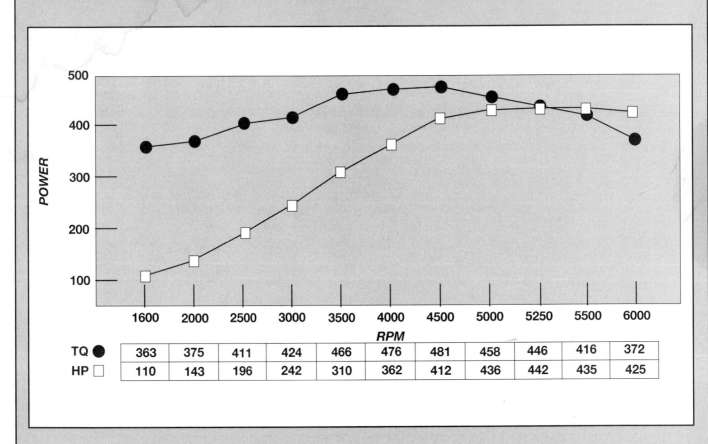

	1600	2000	2500	3000	3500	4000	4500	5000	5250	5500	6000
TQ ●	363	375	411	424	466	476	481	458	446	416	372
HP □	110	143	196	242	310	362	412	436	442	435	425

420 cid EFI
525 HP/540 LBS-FT.

Let's step up to the Godzilla of small-block torque monsters with the 420. The concept here is simple—make maximum torque at the least rpm to make even the heaviest car brutally fast. This is Rat motor displacement in a light, small-block package.

This engine is based on a stroker 4340 steel Callies crankshaft with steel Oliver 5.875-inch rods and forged, flat top custom pistons with 1/16-inch rings and Clevite 77 bearings. This particular engine used a Chevrolet Bow Tie block but this engine could be built using a production 400 block. Actuating the valves is a Comp Cams mechanical roller cam with 236/244 duration at .050-inch tappet lift and .549/.549-inch lift with 1.5:1 Comp Cams roller rockers and a lobe separation angle of 110 degrees.

There are a variety of heads available for this engine that would work well including our ported Corvette heads. This particular engine used a set of Lingenfelter-ported Air Flow Research 190 aluminum castings. The SuperRam base, runners and plenum were also ported and fitted with a Lingenfelter 58mm 1000 cfm billet throttle body. This test was performed using an open set of 1 7/8-inch headers. Controlled by an ACCEL/DFI ECM and using 36-pound-per-hour injectors, this rascal cranked out some truly astounding numbers. Even at a mere 1600 rpm the 420 made over 400 lbs-ft. of torque with peak torque at 4500 rpm and 525 horsepower at 5500 rpm. With its incredibly stout torque curve, this is an engine that demands a big set of sticky tires.

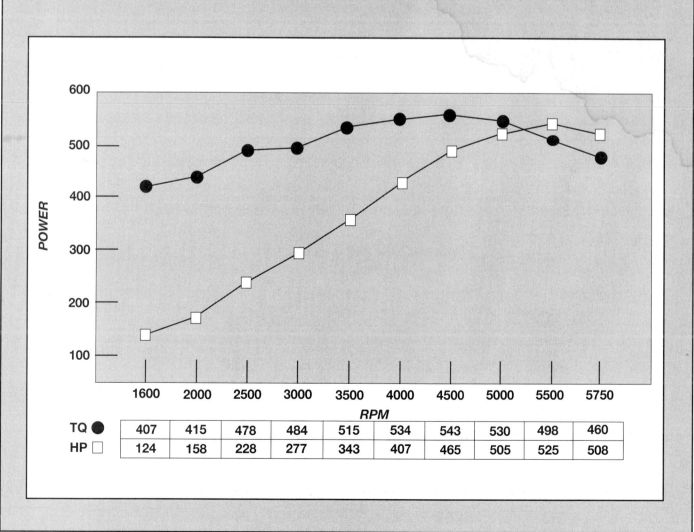

RPM	1600	2000	2500	3000	3500	4000	4500	5000	5500	5750
TQ ●	407	415	478	484	515	534	543	530	498	460
HP □	124	158	228	277	343	407	465	505	525	508

EFI EMISSONS-LEGAL SMALL-BLOCKS

Our EFI engines can be placed into two groups. For the late-model LT1 Corvette and Camaro/Firebird, we offer two emissions-legal engine packages for either the automatic or 6-speed. For the automatic cars, we recommend the SuperRam intake, since it generates more torque to accommodate the wide rpm drop between shifts. For the 6-speed manual-equipped cars, I usually opt for a ported version of the factory LT1 intake. This sacrifices some torque compared to the SuperRam, but makes more horsepower. All of these packages increase power over stock throughout the rpm band. Each of these packages are available in either 350 cid or 383 displacements.

As an example of this emissions-legal power, *Motor Trend* magazine tested a number of emissions-legal performance F-cars and our 383, SuperRam-equipped, 6-speed Firebird was the class of the field that included two supercharged F-cars. Our Firebird ran 13.10's at 113 mph on a very slippery drag strip and tested 410 rear wheel horsepower on a chassis dyno! Subsequent testing at a track with better traction and a set of small slicks generated a stunning 12.10 at over 116 mph! Keep in mind, this is through the complete exhaust including the catalytic converters!

383 cid EFI LT1
425 HP/468 LBS-FT.

This combination uses the Lingenfelter SuperRam intake manifold intended for an automatic transmission combination. We also offer the 383 with a ported factory LT1 intake to be used with the manual 6-speed transmission. With the six-speed, this same combo generates a peak torque of 436 lbs-ft. at 4500 rpm and a peak horsepower of 440 horsepower at 5500 carrying that power through 6000 rpm. This is a higher power peak which works best with the tighter gear spread of the new T-56 6-speed manual transmission.

In either intake configuration, the 383 cid LT1 short block receives some fairly trick parts with the addition of a Lunati 1-piece seal, steel 3.75-inch stroker crankshaft, JE forged pistons and a set of 4340 steel Oliver rods. We change the cam to one of our own hydraulic roller designs with 219/219-degrees of duration at .050 with .503/525-inch lift and a 112-degree lobe separation angle. The LT1 heads are retained, but CNC-ported and fitted with Lingenfelter's 2.00/1.56-inch stainless steel valves. The box is also ported to gain the most advantage from the casting. The following dyno numbers were generated using the factory iron LT1 exhaust manifolds, but without the remaining factory exhaust.

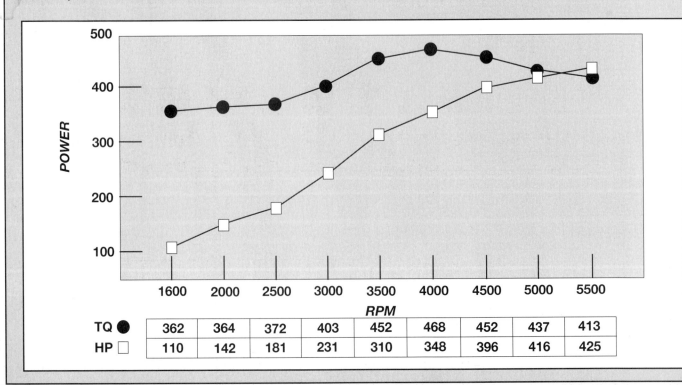

	1600	2000	2500	3000	3500	4000	4500	5000	5500
TQ ●	362	364	372	403	452	468	452	437	413
HP □	110	142	181	231	310	348	396	416	425

350 cid LT5 DOHC
500 HP/425 LBS-FT.

For those of you with a taste for the exotic, we have also worked some magic on the four-cam LT5 engine powerplant for the ZR-1 Corvette. The short block remains basically stock in this configuration with attention paid to creating a blueprinted short block assembly. The power comes mainly from the cylinder head and intake manifold port work. Retaining the stock camshafts, but retiming them for best power, we are able to coax this DOHC engine to an incredible 500 horsepower at 7000 rpm while retaining a stock quality idle! If you're into the ZR-1, we have a plan to make it fly!

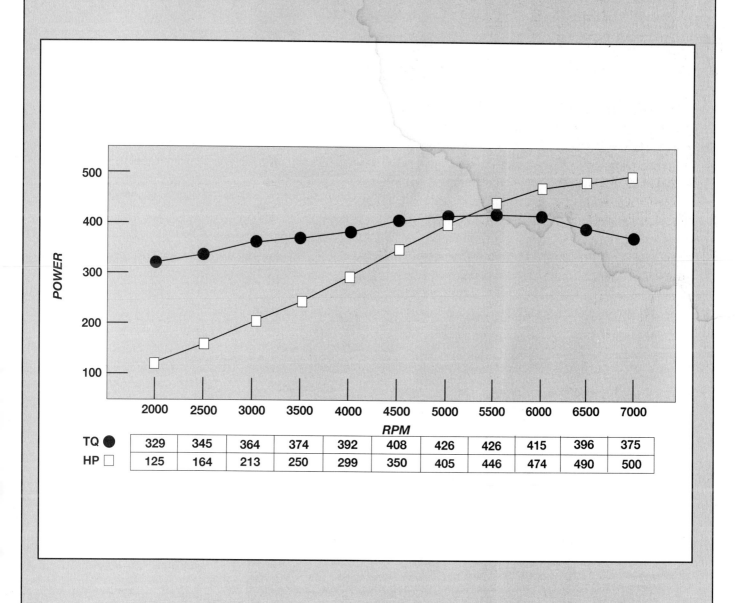

	2000	2500	3000	3500	4000	4500	5000	5500	6000	6500	7000
TQ ●	329	345	364	374	392	408	426	426	415	396	375
HP □	125	164	213	250	299	350	405	446	474	490	500

385 cid LT5 DOHC
570 HP/495 LBS-FT.

This is our latest and most powerful LT5. The increase in displacement is traced to a custom forged steel stroker billet 4340 steel crankshaft that increases the stroke to 3.820 inches while the bore remains at 4.00 inches. Custom cylinder sleeves are used and fitted with forged JE 11.3:1 pistons that accommodate the longer stroke and longer 5.850-inch Oliver 4340 steel rods. As with the previous LT-5, the heads are fully ported. Special titanium retainers are used to enhance the higher rpm potential and are matched to a pair of special profile exhaust camshafts more appropriate to the larger displacement. As you can see, this is a plenty stout package, making over 570 horsepower at 6800 rpm. This will push even a heavy production Corvette solidly into the 11's while retaining its street manners. For those with the stout heart and wallet, this is the killer LT5.

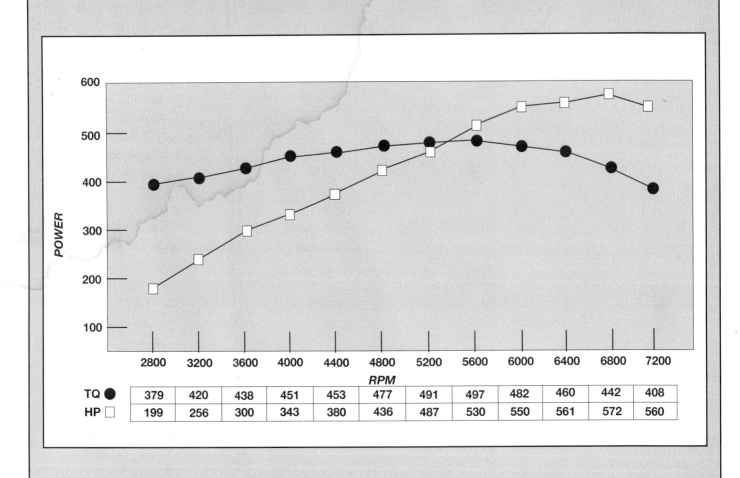

	2800	3200	3600	4000	4400	4800	5200	5600	6000	6400	6800	7200
TQ ●	379	420	438	451	453	477	491	497	482	460	442	408
HP ☐	199	256	300	343	380	436	487	530	550	561	572	560

INDEX

OTHER HP AUTOMOTIVE BOOKS

 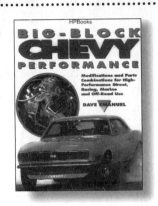 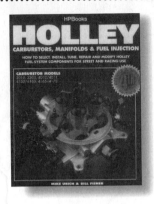 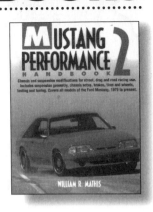

HANDBOOK SERIES

Auto Electrical Handbook
Auto Math Handbook
Automotive Paint Handbook
Baja Bugs & Buggies
Brake Handbook
Camaro Restoration Handbook
Corvette Weekend Projects
Fiberglass & Composite Materials
Metal Fabricator's Handbook
Mustang Restoration Handbook
Mustang Weekend Projects
Off-Roader's Handbook
Paint & Body Handbook
Sheet Metal Handbook
Street Rodder's Handbook
Turbochargers
Turbo Hydra-matic 350
Understanding Automotive Emissions Control
Welder's Handbook

CARBURETORS

Holley 4150
Holley Carburetors, Manifolds & F
Rochester Carburetors
Weber Carburetors

PERFORMANCE SERIE'

Big-Block Chevy Performance
Camaro Performance
Chassis Engineering
How to Hot Rod Big-Block Chev

How to Hot Rod Small-Block Chevys
How to Hot Rod Small-Block Mopar Engines
How to Hot Rod VW Engines
How to Make Your Car Handle
John Lingenfelter On Modifying Small-Block Chevy Engines
Mustang Performance
Mustang Performance 2
1001 High Performance Tech Tips
Race Car Engineering & Mechanics
Small-Block Chevy Performance

REBUILD SERIES

Rebuild Air-Cooled VW Engines
Rebuild Big-Block Chevy Engines
Rebuild Big-Block Ford Engines
Rebuild Big-Block Mopar Engines
Rebuild Small-Block Chevy Engines
Rebuild Small-Block Ford Engines
Rebuild Small-Block Mopar Engines
Rebuild Ford V-8 Engines

GENERAL INTEREST

Auto Dictionary
llector's Handbook
s Funny Cars
GM Muscle Cars

R CALL: 1-800-223-0510

Publishing Group
dison Avenue
ork, NY 10016